CLIMATE CHANGE AND EDUCATION PLAYBOOK

INVESTING IN EDUCATION AND SKILLS FOR CLIMATE RESILIENCE IN ASIA AND THE PACIFIC

NOVEMBER 2024

ASIAN DEVELOPMENT BANK

Contents

Tables, Figures, and Boxes

Foreword

Effectively addressing climate change and enabling transition to low-carbon and climate-resilient economies are among the most pressing challenges faced by governments in the Asia and Pacific region. Governments in the region have joined global efforts to contain the climate threat through nationally determined contributions (NDCs) under the Paris Agreement. As we approach the next update of the NDCs in February 2025, it is timely to recognize the importance of investing in education and skills to enhance capacities to tackle climate change and promote climate resilient economic pathways in a durable manner. The Asian Development Bank (ADB) has put climate financing as a major priority for our investments in the region with the ambition to provide $100 billion in climate finance to its developing member countries (DMCs) in Asia and the Pacific region by 2030. This priority informs the work of all sectors of ADB operations, including education.

The education sector is a critical enabler of timely and sustainable climate action. Education equips people, workers, youth, children and women with skills, knowledge, and attitudes to develop climate-resilient solutions. Investments in education and training are pivotal for decarbonizing transport sectors like rail, maritime, and ports. By equipping professionals with advanced knowledge in sustainable technologies and practices, we can promote adoption of cleaner energy. ADB is spurring extensive work with regional and international partners under the Energy Transition Mechanism (ETM) to accelerate the transition from fossil fuels to clean energy. Specialized training for renewable energy integration, energy-efficient design, adoption of new technologies such as green hydrogen and use of green logistics will help to scale up ETM, a critical development vision for our region. As climate-induced displacements rise, frontline health workers have a critical role in preparing for and responding to climate-related health challenges as first responders and support communities to become climate resilient. For this, they need requisite training and capacity development.

I am delighted that COP29 will host a human development day for the first time and congratulate the excellent work of the Azerbaijan Presidency in this matter. This is a timely global signal for governments to invest strongly in human development for climate mitigation and adaptation.

The human and social development sector group of ADB is launching this climate change and education playbook to provide ADB DMCs with strategies to make the education sector more climate-ready, and thereby more future-ready. The playbook offers practical solutions for climate education in basic and secondary, technical and vocational education and training and higher education, research and innovation. It advocates building climate resilience in education facilities. Extreme weather events exacerbate the learning crisis by damaging schools, resulting in school closures and loss of instructional hours.

The case studies presented in the playbook demonstrate how climate-orientations in education systems can be promoted toward achieving a prosperous, inclusive, resilient, and sustainable Asia and the Pacific.

Ramesh Subramaniam
Director General
Sectors Group
Asian Development Bank

Acknowledgments

Alexander Tsironis, education specialist, Asian Development Bank (ADB), prepared this publication with research and content and contributions from external experts and ADB specialists. The publication was led by Shanti Jagannathan, director, Human and Social Development, ADB with overall guidance from Ayako Inagaki, senior director, Human and Social Development, ADB. We are thankful to Christina Kwauk for her collaboration with ADB in research on global practices and tool kits in integrating climate change and education.

Special thanks are extended to ADB colleagues from various departments for their feedback and contributions to the publication: Sunhwa Lee, principal social sector specialist; Ryotaro Hayashi, senior social sector economist; Per Borjegren, principal education specialist; Arghya Sinha Roy, principal climate change specialist (climate change adaptation); David Morgado, senior energy specialist; Oleksiy Ivaschenko, senior social protection and jobs specialist; Vishal Aditya Potluri, social sector specialist; Namrata Saraogi, social sector specialist; Martino Pelli, senior economist; Zonibel Woods, senior social development specialist (gender and development); Jin Ha Kim, gender specialist (climate change); and Ryan Anthony M. Bestre climate change officer.

Case studies from ADB projects were contributed by Asako Maruyama, principal education specialist; Cindy Bryson, senior social development specialist; Ryotaro Hayashi, senior social sector economist; Unika Shrestha, social sector economist; as well as international experts Sangita Budhathoki, Suraya Acharya, Shaun M. Wellbourne-Wood, and Swetal Sindhvad.

Special gratitude is extended to external reviewers for their valuable feedback: Manuela Prina and Romain Boitard (European Training Foundation); Borhene Chakroun, Hervé Huot-Marchand, and Rika Yorozu (UNESCO); Kenneth Barrientos and Friedrich Huebler (UNESCO-UNEVOC); Hae Kyeung Chu (ILO); Akihiro Fushimi, Maria Melizza Tan, and Jukka Tulivuori (UNICEF); Brigitta Christa Lambertz (GIZ); and B.A. Barendregt, M. Nijemeisland, M.A. Holtkamp, and N. Hopman (Leiden University). Emmanoelle G. Feliciano, climate change consultant, as well as international experts Christina Kwauk and Mahfuzuddin Ahmed provided constructive feedback and hands-on suggestions that significantly added technical depth to the publication.

Support in the finalization of this publication was provided by ADB colleagues Dorothy Geronimo, senior education officer; Evangelyn Medina, ADB consultant; Rodel Bautista, publishing product coordinator, Department of Communications and Knowledge Management; and Maria Theresa Mercado, editorial consultant.

Glossary

Adaptation refers to adjustments to actual or expected climate change and its effects that moderates harm or exploits beneficial opportunities (ADB 2023c).

Climate change education helps people understand and address the impacts of the climate crisis, empowering them with the knowledge, skills, values and attitudes needed to act as agents of change (UNESCO n.d.)

Climate-ready education systems (i) teach transformative climate literacy and green skills across education subsectors; (ii) operate climate-resilient school facilities and operations that have been adapted to climate risks; (iii) operate sustainable and low-carbon school facilities and operations that minimize carbon emission and pollution; and (iv) promote climate-oriented research and development, entrepreneurship, and incubation that foster innovations in climate technologies. Thereby, education systems are adapted to climate change and the human development needs of low-carbon and climate-resilient economies and societies.

Climate resilience refers to the capacity to prepare for, respond to, and recover from the impacts of events caused by climate hazards with minimal damage to social well-being, the economy, and the environment (ADB 2023c).

Climate-resilient schools and operations have the infrastructure as well as institutional and human capacity, such as trained school staff and emergency plans, to withstand, adapt to, and recover from extreme weather events (e.g., heat wave, floods) thereby maintaining learning continuity, a conducive learning environment, and student and school personnel safety.

Climate science education involves teaching and learning about the scientific perspective on the human-caused changes to the climate. It prepares learners to understand the essential principles of the earth's climate system and its interactions with human systems, know how to assess scientifically credible climate information, communicate about climate change in a meaningful way, and make informed and responsible decisions regarding actions that may affect climate (US Global Change Research Program 2024).

Education for climate action teaches the skills needed for jobs in the low-carbon economy and the competencies to understand and respond to the impacts of climate change. It prepares learners to act with self-efficacy and in ways that promote climate resilience, environmental integrity and social responsibility in daily life and the world of work. Education for climate action shares commonalities with other traditions of environmental education such as climate change education and sustainability education, while emphasizing green skills and transformative climate literacy.

Education for sustainable development empowers learners to make informed decisions and take responsible actions for environmental integrity, economic viability, and a just society, for present and future generations, while respecting cultural diversity. It is about lifelong learning and is an integral part of quality education (UNESCO Institute for Statistics n.d.).

Gender-responsive pedagogy refers to teaching and learning processes that pay attention to the specific learning needs of girls and boys. In practical terms, this means that the learning materials, methodologies, content, learning activities, language use, classroom interaction, assessment, and classroom set up are scrutinized to respond to specific needs of boys and girls in the teaching–learning process (UNESCO International Institute for Education Planning n.d.)

Green jobs are decent jobs that contribute to protecting and restoring the environment and addressing climate change. Green jobs can be found in both the production of green products and services, such as renewable energy, and in environmentally friendly processes, such as recycling. Green jobs help improve energy and raw material efficiency, limit greenhouse gas emissions, minimize waste and pollution, protect and restore ecosystems, and support adaptation to the impacts of climate change (UNDP 2023).

Green skills include skills, competencies, knowledge, abilities, values, and attitudes needed to work in low-carbon and climate-resilient economies. They are technical skills required to adapt or implement standards, processes, services, products, and technologies to protect ecosystems and biodiversity, and to reduce energy, materials, and water consumption. They can also include transversal skills linked to sustainable thinking and acting, and basic environmental knowledge relevant to work in all economic sectors and occupations.

A **just transition** means greening the economy in a way that is as fair and inclusive as possible to everyone concerned, creating decent work opportunities and leaving no one behind (ILO 2024).

A **low-carbon and climate-resilient economy** is characterized by economic activities that deliver goods and services that generate significantly lower emissions of greenhouse gases, promote sustainability and are adapted to climate change and its expected impacts.

Nationally determined contributions are climate pledges and action plans that each country is required to develop in line with the Paris Agreement goal of limiting global warming to 1.5° C. They outline mitigation and adaptation priorities that a country will pursue to reduce greenhouse gas emissions, build resilience, and adapt to climate change (UNDP 2023).

Nonstructural measures are measures not involving physical construction that use knowledge, practice, or agreement to reduce risks and impacts, in particular, through policies and laws, public awareness raising, training and education (IUCN 2011).

Mitigation refers to activities and efforts to reduce or limit greenhouse gas production or enhance greenhouse gas sequestration (ADB 2023c).

Sustainable and low-carbon schools and operations are physically constructed and have the institutional and human capacity to minimize carbon emissions and pollution and create a healthy learning environment for students. They can include structural (e.g., installation of solar panels) and nonstructural measures (e.g., sustainable procurement, energy efficiency management).

Transformative climate literacy is a core skill that prepares learners to design solutions to and act with self-efficacy to address climate change challenges in ways that promote environmental integrity and social responsibility. It goes beyond climate awareness and climate science knowledge and includes a breadth of subject-specific (e.g., interpreting climate data), transferable (e.g., anticipating future impacts of climate change), and transformative competencies (e.g., valuing different perspectives and collectively develop climate related solutions that promote equality) that empowers learners to act cooperatively, be flexible, think critically, respect diversity, care for the environment, and be actively involved in finding solutions to climate change.

Transformative learning empowers all learners with the knowledge, skills, values, and attitudes to live cooperatively, be flexible, think critically, respect diversity, care for the environment, and be actively involved in finding solutions.

Climate change and the transition to a low-carbon and climate-resilient economy are the key policy challenges in Asia and the Pacific. Among all regions in the world, Asia and the Pacific faces the largest exposure to climate change and has incurred the greatest economic losses caused by disasters triggered by natural hazards between 1980–2022. Nine out of 15 nations most affected by extreme weather are in Asia and the Pacific. At the same time, the region is increasingly a contributor to the global climate crisis with its share of global greenhouse gas emissions doubling from 22% in 1990 to over 50% in 2024. At the current trajectory, climate targets are not met across the region, which needs to prepare for further exposure to extreme weather events that cause economic disruptions, food and water insecurity and undermines socioeconomic development. Actions are urgently needed to enhance climate resilience and accelerate the transition to low-carbon economies and societies.

Climate change also impacts the education sector. Like all sectors, the education sector is exposed to the risks and opportunities of climate change and requires strategic rethinking. Therefore, the climate change and education playbook examines the implications of climate change on education systems in Asia and the Pacific. It explores how education is affected by—and therefore needs to adapt to—climate risks and how the sector can be an important enabler of the transition to a low-carbon and climate-resilient economy as set out in the nationally determined contributions (NDCs) and the Sustainable Development Goals (SDG) including SDG 4 Quality Education, SDG 9 Industry, Innovation, and Infrastructure, SDG 12 Responsible Consumption and Production, and SDG 13 Climate Action, as well as, SDG 6 Clean Water and Sanitation, SDG 7 Affordable Clean Energy, SDG 11 Sustainable Cities and Communities, SDG 14 Life Below Water, and SDG 15 Life on Land. The playbook provides education policymakers and project developers with a call to action, action principles, and seven case studies for rethinking education systems making them climate ready.

Implications of Climate Change and the Low-Carbon and Climate-Resilient Economy and Society on the Education Sector

Climate change exacerbates the learning crisis. Extreme weather events, such as extreme heat, floods, and storms, negatively affect learning outcomes. Extreme heat and other calamities caused 32 days of school closures in the Philippines in the school year 2023–2024, disrupting learning continuity. In Pakistan, the floods of 2022 damaged or destroyed at least 17,205 schools, disrupting the education of approximately 2.6 million children. In India, students who experience prolonged exposure to storms are more likely to experience educational delays, are less likely to complete higher education, and are less likely to secure salaried employment. Overall, climate change negatively impacts learning outcomes, affecting schools and children that do not have adaptive capacities the most. Making education facilities and operations climate resilient is a growing policy priority in the education sector to prevent learning losses, physical harm to learners and teachers, and damage to school assets that prolong school closures. Schooling is of critical importance as an additional year of education has been linked with increases in pro-climate beliefs and behaviors.

Labor markets and society face extensive needs for green skills and education for climate action (E4CA). With the introduction of climate technologies and sustainable business practices across industries, new job profiles and skill needs are emerging in labor markets. In 2021, global clean energy employment surpassed that of fossil fuels, reaching a total of 35 million jobs. In India, solar energy capacity installed has increased by almost 1,800% between 2015 and 2021 and the industry is projected to employ more than 3 million workers by 2030. In Viet Nam, the market share of electric two-wheelers, many of which are locally assembled, has doubled in 3 years from 5.4% in 2019 to 10% in 2021. Globally, green patent development has increased by 100% over the past 10 years, resulting in an increase in investment and economic growth. It is estimated that the transition to a low-carbon economy can create up to 232 million jobs by 2030 in Asia and the Pacific. This creates opportunities for women to enter nontraditional jobs (e.g., in the construction or energy sector) and emphasizes the need for a just transition that leaves no one behind, including 8.3 million workers in the coal industry. In addition, the 4.7 billion citizens in the region require the knowledge, attitudes, and skills to live sustainable lives that are informed by environmental integrity and climate awareness. Overall, there is a significant need and demand for E4CA and green skills.

Education systems and NDCs have not yet caught up with the human development needs of low-carbon and climate-resilient economies and societies. Most education systems do not comprehensively teach the needed green skills and climate literacy yet. Around 75% of national curricular frameworks in Asia and the Pacific have no or only minimal focus on climate change. Nearly one in four youth in the region reports either not knowing anything about climate change or knowing something about climate change, but not being capable of explaining what it is. Globally, the demand for green jobs is outpacing the supply of green workers. In addition, E4CA and skills development are not substantially included in many NDCs and disconnected from countries' climate action plans. As a result, the potential of education systems as a lever to achieve climate targets and enable green jobs, climate resilience and action is underutilized.

Making Education Systems Climate-Ready Is a Key Policy Agenda

Making education systems climate-ready is a key policy agenda in the coming years. Education systems need to holistically adapt to climate risks and the skill needs of low-carbon and climate-resilient economies and societies. They need to become climate-ready by (i) teaching transformative climate literacy and green skills across all education subsectors; (ii) adapting school facilities and operations to climate risks toward climate resilience, including disaster risk preparedness; (iii) operating sustainable and low-carbon school facilities and operations; and (iv) promoting climate-oriented research and development, entrepreneurship, and incubations services. These four points form the key action areas for investing in education for climate resilience in the future.

Climate-ready education systems help overcome the climate and learning crisis, creating a virtuous cycle. Climate-ready education systems can have far-reaching impacts that not only enables the transition to the low-carbon and climate-resilient economy but also fundamentally reinforces and safeguards learning outcomes through various pathways that start with E4CA. For example, E4CA enhances the climate resilience of communities, resulting in less disruptions to livelihoods, which in turn improves the school attendance of children. Green skills enable green jobs, and green jobs contribute to lowering carbon emissions, which reduces climate impacts on schools—preventing school closures and averting potential learning losses. Climate-resilient school facilities and operations withstand climate impacts and ensure that instructional hours are maintained despite extreme weather events. Therefore, making education systems climate-ready is also about safeguarding learning outcomes and preventing an exacerbation of the learning crisis.

Key Directions for Climate-Ready Education Systems

Integrate E4CA in each education subsector. Each education subsector focuses on developing different aspects of E4CA. Basic and secondary education focuses on transformative climate literacy, technical and vocational education and training (TVET) on green skills for green jobs, and higher education on advanced climate knowledge across various disciplines. In addition, higher education also needs to support climate-oriented research and development that promotes climate innovation and the adoption and localization of climate technologies and solutions.

Take a holistic approach when introducing E4CA. First, E4CA needs to be holistically integrated into education systems, starting with integrating climate-related learning goals into regulatory education instruments such as national curriculum frameworks. Second, building climate-ready education systems requires both infrastructure as well as capacity development and adjustments throughout the education system and in schools. This includes capacity development of teachers, principals, and facility managers, the construction of climate-resilient facilities, and updating of teaching, learning, and assessment materials. The whole-of-school approach provides a guiding framework. Third, E4CA must be inclusive and gender-transformative from the outset to ensure equitable access and participation in E4CA for women and girls.

Introduce transformative climate literacy in basic and secondary education. Transformative climate literacy requires a range of competencies that include understanding climate interactions, addressing climate risks, and promoting equity and justice. Teaching climate literacy is therefore not limited to climate science but requires transformative learning, which shifts students' perspectives and behaviors toward sustainability, climate resilience, and fairness for people and communities who are disproportionately affected by climate-related challenges. Given the diverse and transformative climate learning outcomes, E4CA should be integrated into the curricula across various subjects including social science and needs to be taught by using place-based, action-oriented, and intersectional teaching methods. Science, technology, engineering, and mathematics (STEM) is a crucial pillar of climate literacy, but also needs to be climate-oriented to allow learners not just to explain the impact of human interactions with Earth's systems, but also to make informed decisions, and develop the agency to provide solutions to socio-ecological crises. The effective introduction of climate literacy also hinges on teacher professional development.

Update TVET qualifications and training programs strategically in line with industry needs.
The transition to a low-carbon economy changes the skills requirements in labor markets. As industries, not just in the clean energy sector, are adopting different climate technologies (e.g., electric vehicles, battery storage, green hydrogen, etc.) and environmental principles (e.g., resource efficiency, circularity, sustainable use, and biodiversity), green skills needs are diverse—resulting in a variety of emerging green job profiles. They include specialist jobs associated with specific green technologies such as solar panel technician; sector-agnostic green jobs such as sustainability manager; and conventional jobs that are supplemented with green skills, such as sustainable urban planners. Besides green specialists, all workers in the economy

need to have basic environmental knowledge to perform jobs and tasks resource-efficiently without causing environmental damage. As a result, qualification standards and training programs need to be updated in targeted manner reflecting the diverse green skills needs and job profiles. Learner-centered teaching, work-based learning, soft skills, entrepreneurship skills, labor market forecasting, and cooperation with business and industry remain important approaches and themes in TVET that require continued capacity development efforts.

Introduce interdisciplinary degree programs and E4CA across disciplines in higher education.
Higher education plays a critical role in offering degree programs that provide skills for highly skilled green jobs in the service sector such as climate finance, environmental policy, teacher education, sustainable urban planning, environmental law, and sustainable business management as well as the technology sector such as engineering and science. Universities can leverage their flexibility in academic structure to integrate E4CA into degree programs in various ways such as integrating climate content into existing courses, creating specialized programs, establishing interdisciplinary degrees, or making climate literacy part of the mandatory core curriculum. Universities are well-positioned to provide interdisciplinary degree programs that can teach the most complex climate related competencies. To support these efforts, universities need to develop organizational structures that foster interdisciplinary collaboration and partnerships with industry and communities to advance climate knowledge and solutions. Community engagement is vital to ensuring university resources are leveraged to address local climate challenges and advance community-led climate adaptation.

Make schools climate-resilient—ensuring education continuity and preventing learning loss.
Climate change poses a significant threat to continuous education delivery and learning outcomes. Each school faces unique climate risks that need to be identified, assessed, and addressed through adaptation measures, making schools climate resilient. Adaptation measures help reduce the impacts of climate risk, allow schools to continue functioning with minimum disruption, and create a culture of safety and resilience. These can include structural measures, such as upgrading school buildings to withstand extreme weather events, as well as nonstructural measures, such as disaster preparedness planning, safety drills, and enabling remote, digital learning. In addition, extreme weather events can result in economic shocks to households, who withdraw their children from schools as a coping mechanism, resulting in lower educational outcomes. Intersectoral solutions that draw on social protection programs can enhance the climate resilience of those households' providing livelihood and financial assistance enabling the continued schooling of children.

Promote sustainable and low-carbon school facilities for environmental, economic, and educational benefits. Schools can contribute to national climate targets, in line with NDC building sector measures by implementing energy-efficient building designs, resource-efficient operations and school management, and renewable energy solutions. These measures not only reduce the environmental impact of school operations but also provide economic and educational benefits, such as reduced operational costs and improved student well-being. Potential measures include the installation of on-site renewable energy generation to improve electricity stability needed to reliably run digital education and green building certification for newly constructed facilities. Energy audits are critical to identify resource-efficiency and climate mitigation opportunities in schools. Sustainable school facilities can be leveraged as living learning labs that empower students to engage in climate action initiatives outside the classroom.

Drive climate innovation through research and development and innovation hubs. Innovation hubs provide innovation, knowledge, and commercialization support services to researchers and industry, which need to be climate oriented in line with countries decarbonization pathways outlined in NDC, national climate action plans, and other industry strategies. This can range from promoting green entrepreneurship for self-employment, supporting local industries in the adoption of climate technologies, developing and registering green patents for licensing to industries, to supporting industries to move into green industry value chains for industrial development. To establish effective and climate-oriented innovation hubs, universities must align the services of such hubs with regional industry needs and develop new capabilities such as industry-oriented research and development, entrepreneurship training, and incubation support. To drive climate innovation, it is key that these services are climate-oriented by, for example, setting up fabrication labs that include climate-related equipment, partnering with green industries, focusing on climate related industry challenges, and building up internal climate related know-how.

Call to Action for Climate-Ready Education Systems

The Asian Development Bank (ADB) aims to support its developing member countries in Asia and the Pacific in accelerating the transition to climate-ready education systems by adopting comprehensive measures that promote climate resilience, skills for green jobs, and climate literacy. ADB aims to support climate mitigation and adaptation investments in education and skills projects and programs to (i) embed education for climate action (E4CA) in curriculum, teaching and learning practices, and education infrastructure; (ii) support policy developments to amplify the role of education in national policy instruments, including the nationally determined contributions (NDCs); (iii) support high quality analytical and knowledge work that fuels new investments in E4CA and climate resilience; and (iv) strengthen partnerships and collaborations with key agencies to advocate common policies and practices while also strengthening actions at the country level. To this end, ADB proposes the following eight calls to action (Figure 1) to be jointly supported by international organizations in support of countries' climate pledges, national actions plans, and the Sustainable Development Goals.

- **Call to Action 1 – Transformative Climate Literacy.** Ensure that in all new basic and secondary education projects, basic climate literacy for all and transformative and gender-responsive E4CA are incorporated in the majority of curricula.

- **Call to Action 2 – Green Skills.** Ensure that in all new technical and vocational education and training, skills, and workforce development projects, green and climate resilience skills are introduced in occupational qualifications, linked to key economic sectors in a country's priority industries, and aligned with NDCs, national climate action plans, and economic development plans.

- **Call to Action 3 – Climate-Oriented Advanced Degree Programs.** Ensure that in all new higher education projects, degree and post-graduate programs are updated or newly developed to include interdisciplinary climate change knowledge and competencies, in alignment with NDCs and national climate action plans.

- **Call to Action 4 - Adaptation.** Ensure that all new investments from 2025 in school and education institution infrastructure and operations are adapted to local climate and disasters risks and meet related building standards.

- **Call to Action 5 - Mitigation.** Ensure that all new investments from 2025 in school and education institution infrastructure, including newly built schools and renovations, are designed and constructed for resource efficiency, including through renewable energy sources, in line with national building codes and green building standards.

- **Call to Action 6 – Climate Innovation.** Ensure that in all new projects that support research, innovation, and start up incubation, the services of research and development and innovations hubs are climate-oriented, in alignment with NDCs, national climate action plans, and other national economic and innovation strategies.

- **Call to Action 7 – Nationally Determined Contributions.** Ensure that all upcoming NDC updates and national climate action plans incorporate substantive measures for harnessing the role of education for climate mitigation and adaptation through appropriate investments in human capital development.

- **Call to Action 8 – Monitoring and Evaluation.** Ensure that in all new education projects, measures and metrics are incorporated that track and monitor E4CA and related capacity and skill development of beneficiaries—including girls, women, and disadvantaged youth.

Guiding principles to inform the eight calls to action. Future investments in education need to be informed by guiding principles to address the complexity and challenges of climate change collaboratively, based on evidence and make cooperation and digital technologies part of solutions in the education sector:

1. **Mobilize (innovative) finance.** Building climate ready education systems requires investments in human and institutional capacities and school infrastructure. Existing climate finance mechanisms need to expand their focus to fund the development climate-ready education system in acknowledgment of the education sector's transformative impact on youth, women, workers, communities, and innovation systems as well as its critical role in the just transition and enabling NDCs.

2. **Cooperate with business and industry.** Business and industry can be key partners in climate-ready education systems. For example, industries are key partners for introducing and updating green qualification standards; in setting up industry-focused, challenge-based research grants; and for running mentorship programs for green entrepreneurs.

3. **Develop intersectoral solutions.** Intersectoral approaches can address the complex problems of climate change. For example, coordinating with social protection line ministries to target vulnerable households with school children can improve children's learning outcomes and coordinating with various line ministries and climate change committees when updating NDCs and national climate action plans.

4. **Promote gender-responsive actions.** All actions must be inclusive and gender-transformative in outcomes, for example, by ensuring gender-responsive school infrastructure designs and promoting female participation and access to science, technology, engineering, and mathematics education and green jobs.

5. **Leverage digital technologies.** Digital technologies can be leveraged to support the E4CA. For example, an online learning platform enhances the climate resilience of education delivery by enabling learning continuity during school closures, or promoting digital skills can support green jobs such as smart metering and irrigation automation.

6. **Conduct high-quality analytics.** Analytics and assessments are critical to fuel new nationally appropriate and tailored investments in E4CA and resilience. Key analytics include green labor market forecast, energy audits and climate risk assessment of schools, and strategic green industry and value chain assessments.

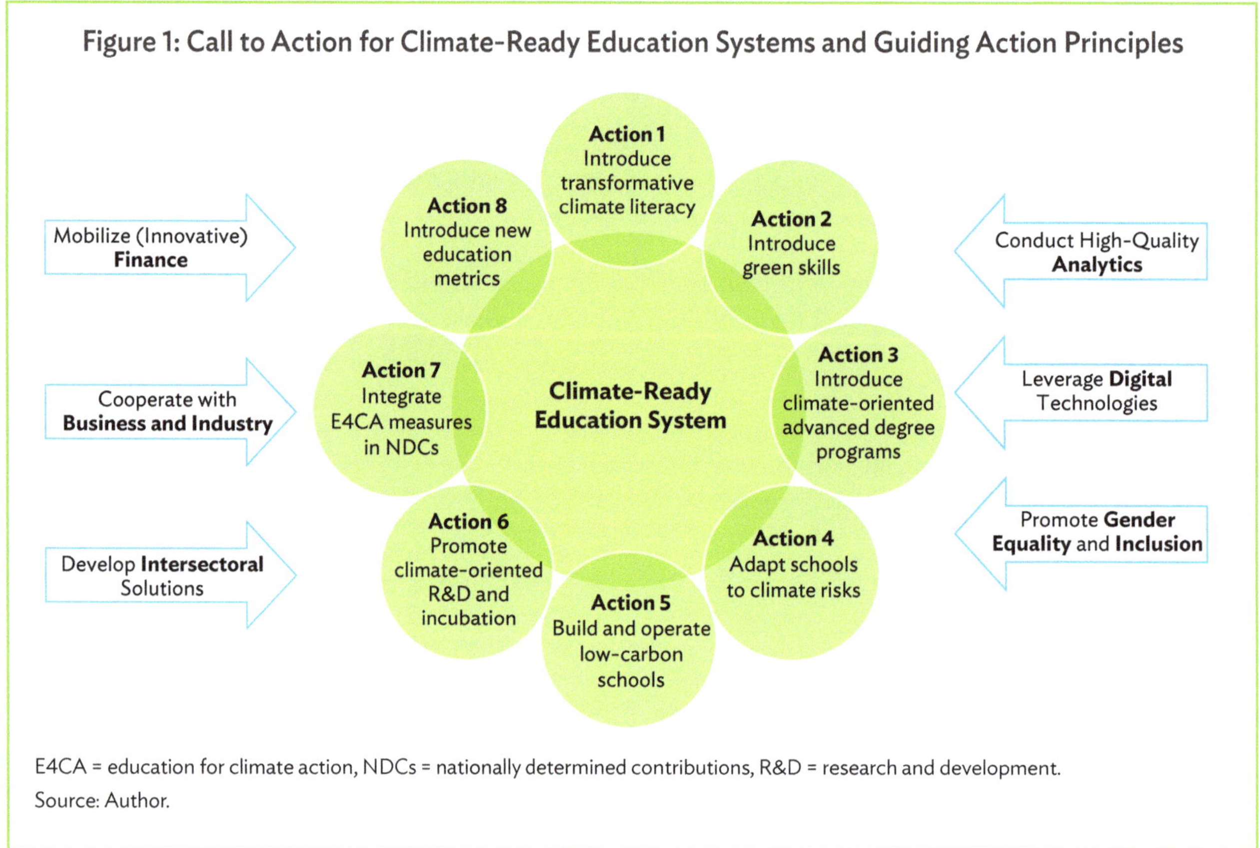

Figure 1: Call to Action for Climate-Ready Education Systems and Guiding Action Principles

E4CA = education for climate action, NDCs = nationally determined contributions, R&D = research and development.
Source: Author.

Abbreviations

ADB	Asian Development Bank
ASU	Assam Skill University
CECCS	Center of Excellence in Climate Change Studies
CSSF	Comprehensive School Safety Framework
DRS	disaster resilience of schools
E4CA	Education for Climate Action
EMIS	education management information system
GDP	gross domestic product
GHG	greenhouse gas
GZAR	Guangxi Zhuang Autonomous Region
HVAC	heating, ventilation, and air conditioning
IF-CAP	Innovative Finance Facility for Climate in Asia and the Pacific Financing Partnership Facility
NAM	national actions for mitigation
NAP	national adaptation plan
NDC	nationally determined contribution
NTU	Nanyang Technological University
PRC	People's Republic of China
PRIMESTeP	Promoting Research and Innovation through Modern and Efficient Science and Technology Parks
R&D	research and development
STEM	science, technology, engineering, and mathematics
STP	science and technology park
TVET	technical and vocational education and training
WSA	whole-of-school approach
UNFCCC	United Nations Framework Convention on Climate Change

The Climate Challenge for Education Systems

Climate change and the transition to a low-carbon and climate-resilient economy are key policy challenges in Asia and Pacific. Among all regions in the world, Asia and the Pacific has the largest exposure to climate change and has incurred the greatest economic losses between 1980–2022 caused by disasters triggered by natural hazards (Dabla-Norris et al. 2023). Nine out of 15 nations most affected by extreme weather are in Asia and the Pacific (IFHV 2023). With the region falling short of its climate targets, extreme weather events are expected to intensify, leading to economic harm and hindering socioeconomic progress (ADB 2024a; ADB 2024d). At the same time, the region is increasingly a contributor to the global climate crisis with its share of global greenhouse gas (GHG) emissions doubling from 22% in 1990 to over 50% in 2024 (ADB 2023, ADB 2024b). Consequently, transitioning to a low-carbon and climate-resilient economy and society is urgently needed to meet climate targets and prevent a global crisis driven by disasters, weather extremes, food and water insecurity, and economic disruption. Like all sectors, the education sector is exposed to the risks and opportunities of climate change and requires strategic rethinking (Reimers 2021). Education systems need to adapt to climate risks becoming climate-resilient and provide learners, including youth, workers, and communities with the competencies and skills needed to live and work in low-carbon and climate-resilient economies and societies.

Climate change exacerbates the learning crisis. Extreme weather events, such as extreme heat, floods, and storms, negatively affect learning outcomes. Schools increasingly face weather-related disruptions to educational services, with school closures being a key cause of learning losses (Saavedra and Sherburne-Benz 2022). Extreme heat and other calamities caused 32 days of school closures in the Philippines in the school year 2023–2024, disrupting learning continuity (PIDS 2024). It has been shown that students who experience prolonged exposure to storms are more likely to experience educational delays, are less likely to complete higher education, and are less likely to secure salaried employment (Pelli and Tschopp 2024). In Pakistan, the floods of 2022 damaged or destroyed at least 17,205 schools, disrupting the education of approximately 2.6 million children (Ministry of Planning Development & Special Initiatives 2022). Making education facilities and operations climate resilient is a growing policy priority in the education sector to prevent learning losses, physical harm to learners and teachers, and damage to school assets that prolong school closures. At the same time, school facilities and operations can lower their own environmental impact to mitigate climate impacts through resource-efficient school facility designs and management.

The transition to a low-carbon economy places new skill needs on education systems. With the introduction of climate technologies and sustainable business practices across industries, new job profiles and skill needs are emerging. In 2021, global clean energy employment surpassed that of fossil fuels, reaching a total of 35 million jobs (IEA 2023). In India, solar energy capacity installed has increased by almost 1,800% between 2015 and 2021 and the industry is projected to employ more than 3 million workers by 2030 (Skill Council for Green Jobs 2023; Tyagi et al. 2022). In Viet Nam, the market share of electric two-wheelers, many of which are locally assembled, has doubled in 3 years from 5.4% in 2019 to 10% in 2021 (International Climate Initiative and UNEP 2023). Globally, green patent development has increased by 100% over the past 10 years, resulting in an increase of investment and gross domestic product (GDP) growth (Lavopa and Menéndez 2023; BBVA Research 2024). Overall, it is estimated that low-carbon economies can create up to 232 million jobs by 2030, assuming that needed green investments are made across the food, land, and ocean use; infrastructure and built environment; and energy and extractives industries (Temasek and Ecosperity 2021). Education systems need to respond to the emerging workforce needs in the low-carbon economy.

The transition to a low-carbon economy creates challenges for communities and workers dependent on carbon-intensive industries. While it is projected that green transition has a net positive impact on job creation (IEA 2022a; Vivid Economics 2021), the fear of job losses is a challenge in accelerating the phase out emission-intensive industries such as the coal industry (Dabla-Norris et al. 2023). People and communities whose livelihoods depend on emission-intensive industries and that are most affected by decarbonization regulations require support to successfully transition into new jobs and livelihoods. This includes 8.3 million workers in the coal industry across Asia and the Pacific (IEA 2022a). Re- and upskilling will be one of the critical policy elements of the just transition to provide workers and communities dependent on emission-intensive industries with the skills needed to participate in and benefit from the transition to the low-carbon economy.

Basic climate knowledge and awareness is required by all people in the region. Beyond green jobs, climate change and the need for mitigation and adaptation actions have resulted in a paradigm shift in human capacity needs in society. Current knowledge, values, and competencies geared toward the needs of carbon-based societies and have resulted in environmental degradation and systemic inequality need to be replaced by transformative skills that enable climate action. The daily actions and habits of all 4.7 billion citizens in Asia and the Pacific need to be informed by environmental integrity, climate resilience, as well as equality as a consideration for a just transition (UNESCAP 2023). Education systems need to prepare learners for this paradigm shift ensuring that the millions of secondary and higher education students who graduate each year possess the needed transformative climate literacy[1] to live and work in a low-carbon and climate-resilient economy and society. Schooling is of critical importance to build climate competencies as an additional year of education has been linked with increases in pro-climate beliefs, behaviors, and most policy preferences (Angrist et al. 2024).

---

[1]   There is no universally agreed definition of transformative climate literacy, the definition used in this publication, see glossary, reflects an emerging consensus that climate literacy includes climate science, needs to be action-oriented, encourage climate-friendly behaviors and attitudes, and consider the equality and justice dimension of climate change.

The Current State of Education for Climate Action

Education systems have not yet caught up with the human development needs of a low-carbon and climate-resilient economy and society. Most education systems are not comprehensively teaching the needed green skills and climate literacy yet. Around 75% of national curricular frameworks in Asia and the Pacific[2] have no or only minimal focus on climate change (UNESCO 2021). Existing teaching approaches to education for climate action (E4CA) tend to overly focus on climate science only, overlooking the transferable, transformative, and interpersonal skills required to change people's perspectives and behaviors for climate resilience and sustainability (GEM Report and MECCE 2024, Reimers 2021). Nearly one in four youth in the region reports either not knowing anything about climate change or knowing something about climate change, but not being capable of explaining what it is (UNESCO 2022), while in a survey, 77% of youth in Asia and the Pacific indicate that they aspire to work in the green economy (Accenture 2022). In another survey, 60% of youth in Central Asia stated that they learn about green issues from social media, compared to 21% from schools (European Training Foundation and UNICEF 2021). Globally, the demand for green jobs is outpacing the supply of green workers (LinkedIn 2023). Energy companies are already reporting difficulties hiring installation and repair jobs, a key bottleneck in the renewable energy labor market (IEA 2023).

Education is insufficiently reflected in nationally determined contributions (NDCs) and disconnected from countries' climate plans. Countries in Asia and the Pacific have outlined mitigation and adaptation priorities to reduce greenhouse gas emissions and adapt to climate change in their NDCs and national sector climate action plans. While the importance of education for climate change is highlighted in the United Nations Framework Convention on Climate Change (UNFCCC) Article 6, education and skills development tends to be insufficiently reflected in NDCs and sectoral climate action plans (Kwauk 2021; UNESCO-UNEVOC 2021). Education and training measures need to be part of NDCs in a substantial way to ensure that education systems strategically enable a country's decarbonization pathway (Figure 2). For education systems to enable a country's decarbonization pathway they need to provide the climate literacy, green skills, as well as research and development (R&D) activities needed to achieve a country's mitigation and adaptation goals and not only teach knowledge and skills that are geared toward carbon-intensive activities,[3] and operate school facilities and operations that contribute to climate mitigation, do not rely on the use of fossil fuel, and are adapted to climate risks (Table 1).

CALL TO ACTION—NATIONALLY DETERMINED CONTRIBUTIONS

Ensure that all upcoming NDC updates and national climate action plans incorporate substantive measures for harnessing the role of education for climate mitigation and adaptation through appropriate investments in human capital development.

---

[2] Countries in the survey included Azerbaijan, Bangladesh, Bhutan, Cambodia, the People's Republic of China, the Cook Islands, India, Indonesia, Japan, Kazakhstan, the Kyrgyz Republic, the Lao People's Democratic Republic, Malaysia, Maldives, Mongolia, Myanmar, Nauru, Nepal, Pakistan, Papua New Guinea, the Philippines, the Republic of Korea, Sri Lanka, Thailand, Tuvalu, and Viet Nam.

[3] The Joint Multilateral Development Banks' (MDB) Methodological Principles for Assessment of Paris Agreement Alignment highlight that investment lending should be consistent with a country's low-carbon, climate-resilient development pathways and shall not lock-in carbon-intensive patterns or the exacerbation of climate risks. This principle can also guide education investments.

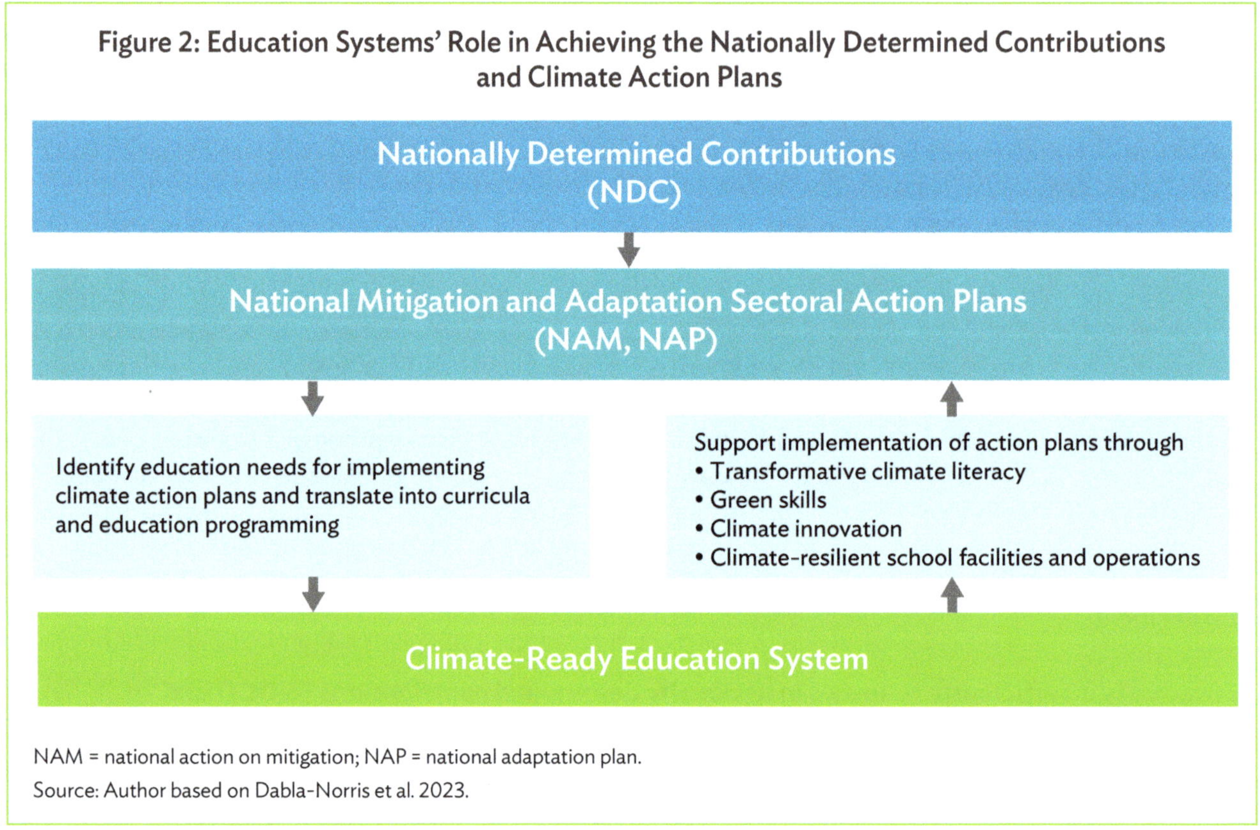

NAM = national action on mitigation; NAP = national adaptation plan.
Source: Author based on Dabla-Norris et al. 2023.

A Policy Agenda for Climate-Ready Education Systems

There are four action areas to make education systems climate-ready. Overall, the two mega trends—climate change and the transition to a low-carbon and climate-resilient economy and society—put new demands on education systems, which need to respond by becoming climate ready. A climate-ready education system (i) teaches transformative climate literacy and green skills, (ii) operates school facilities and operations that are adapted to climate risks, (iii) operate sustainable and low-carbon school facilities and operations, and (iv) promote climate-oriented research and development, entrepreneurship, and incubation (Figure 3). These four points form the key action areas for making education systems and schools climate-ready.

Climate-ready education systems can help overcome the climate and the learning crisis. Climate-ready education systems can have far-reaching impacts that not only enables the transition to the low-carbon and climate-resilient economy but also fundamentally reinforces and safeguards learning outcomes (Figure 4). For example, E4CA creates climate literacy and green skills that provide the skills needed for green jobs and sustainable living. Green jobs and sustainable living in turn result in positive environmental, economic, health and social impacts, which in turn can improve school attendance and children's readiness to learn, thereby reinforcing learning outcomes. In addition, climate adaptation measures create climate-resilient schools that are prepared to provide education services despite disruptive extreme weather events ensuring learning continuity, which again ensures learning outcomes. When these activities are gender-sensitive, outcomes will be gender-transformative, improving learning outcomes as well as access to jobs of women and girls.

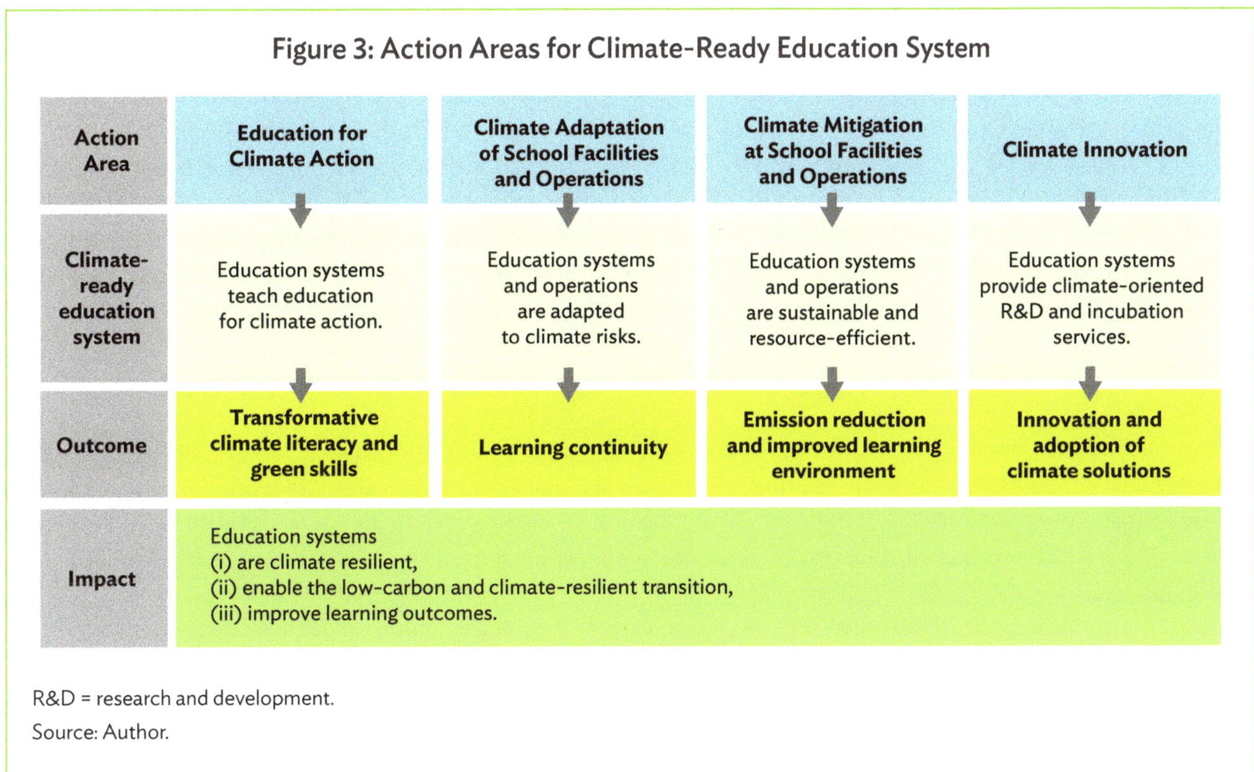

Figure 3: Action Areas for Climate-Ready Education System

R&D = research and development.
Source: Author.

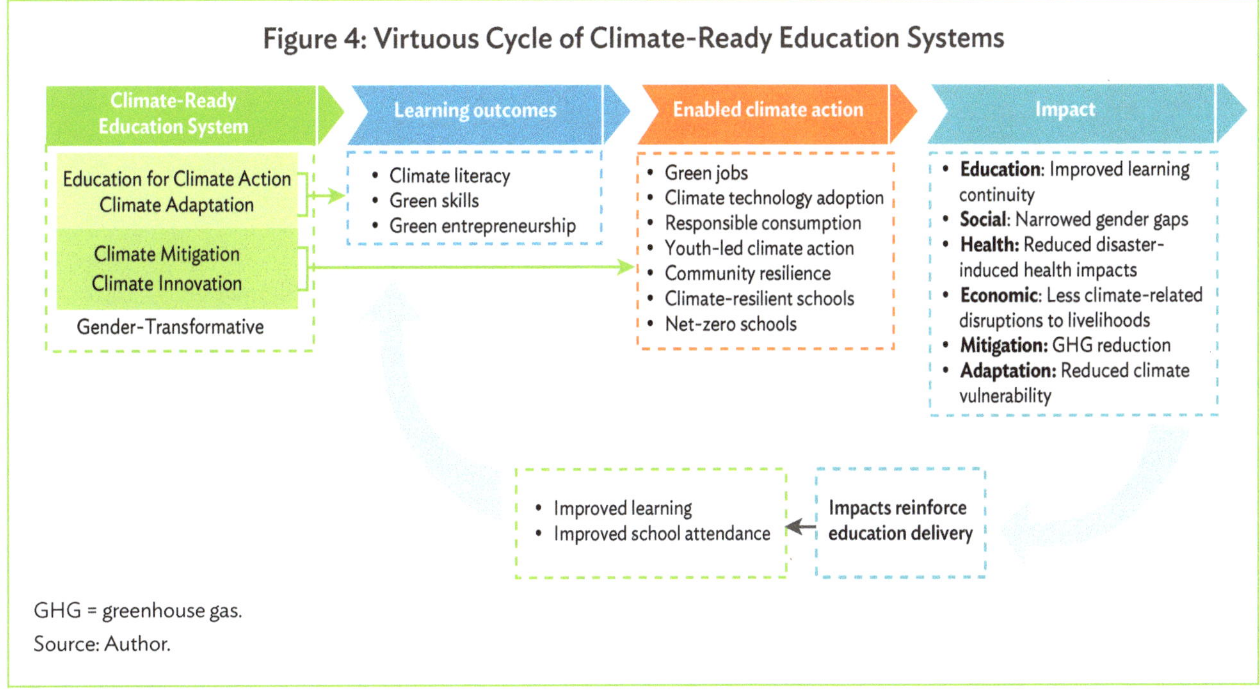

Figure 4: Virtuous Cycle of Climate-Ready Education Systems

GHG = greenhouse gas.
Source: Author.

Climate-ready education systems are key to achieving the Sustainable Development Goals (SDGs).
The virtuous cycle (Figure 4) highlights that education is a strategic enabler for achieving a range of SDGs
and targets, with an explicit relevance for SDG 4 Quality Education, SDG 5 Gender Equality, SDG 9 Industry,
Innovation, and Infrastructure, SDG 12 Responsible Consumption and Production, and SDG 13 Climate
Action (Table 1). While also enabling SDG 6 Clean Water and Sanitation, SDG 7 Affordable Clean Energy,
SDG 11 Sustainable Cities and Communities, SDG 14 Life Below Water, and SDG 15 Life on Land. The nexus
of E4CA with improved learning outcomes, innovation, responsible consumption, jobs, and climate action
(mitigation and adaptation) underlines that investing into climate ready education system needs to receive
greater attention to achieve climate targets of NDCs as well as the various SDGs.

Table 1: Selected Sustainable Development Goals Enabled Through Climate-Ready Education Systems

SDG Goals	SDG Targets
4 QUALITY EDUCATION	SDG 4.1. By 2030, ensure that all girls and boys complete free, equitable and quality primary and secondary education leading to relevant and effective learning outcomes
	SDG 4.4. By 2030, substantially increase the number of youth and adults who have relevant skills, including technical and vocational skills, for employment, decent jobs and entrepreneurship
	SDG 4.7. By 2030, ensure that all learners acquire the knowledge and skills needed to promote sustainable development, including, among others, through education for sustainable development and sustainable lifestyles, human rights, gender equality, promotion of a culture of peace and non-violence, global citizenship and appreciation of cultural diversity and of culture's contribution to sustainable development
	SDG 4.a. Build and upgrade education facilities that are child, disability and gender sensitive and provide safe, non-violent, inclusive and effective learning environments for all
5 GENDER EQUALITY	SDG 5.7. Enhance the use of enabling technology, in particular information and communications technology, to promote the empowerment of women
9 INDUSTRY, INNOVATION AND INFRASTRUCTURE	SDG 9.5. Enhance scientific research, upgrade the technological capabilities of industrial sectors in all countries, in particular developing countries, including, by 2030, encouraging innovation and substantially increasing the number of research and development workers per 1 million people and public and private research and development spending
12 RESPONSIBLE CONSUMPTION AND PRODUCTION	SDG 12.8. By 2030, ensure that people everywhere have the relevant information and awareness for sustainable development and lifestyles in harmony with nature
	SDG 12.a. Support developing countries to strengthen their scientific and technological capacity to move towards more sustainable patterns of consumption and production
13 CLIMATE ACTION	SDG 13.1. Strengthen resilience and adaptive capacity to climate related disasters
	SDG 13.3. Improve education, awareness-raising and human and institutional capacity on climate change mitigation, adaptation, impact reduction and early warning
14 LIFE BELOW WATER	SDG14.a. Increase scientific knowledge, develop research capacity and transfer marine technology, taking into account the Intergovernmental Oceanographic Commission Criteria and Guidelines on the Transfer of Marine Technology, in order to improve ocean health and to enhance the contribution of marine biodiversity to the development of developing countries, in particular small island developing States and least developed countries

SDG = Sustainable Development Goal.
Source: Author.

The *Climate Change and Education Playbook* guides future investments in education. The playbook outlines the strategies and actions that education policymakers and project developers can use to make education systems climate ready. Section 2 outlines in detail key directions and strategies across the four action areas (Figure 3), for (i) introducing E4CA in basic and secondary, technical and vocational education and training (TVET), and higher education, (ii) implementing climate adaptation measures to make school facilities climate resilient, (iii) implementing climate mitigation measures to reduce the carbon footprint of school operations, and (iv) delivering climate-oriented R&D and incubation services to promote climate innovation. Section 3 summarizes the strategies into eight calls to action and guiding principles that education policymakers and project developers need to consider in future education investments and interventions. To inspire future actions, section 4 outlines seven case studies that illustrate how education systems in Asia and the Pacific are already transforming becoming climate ready.

A student walks past the Burgos Wind Farm in Ilocos Norte, Philippines.

2 Key Directions for Climate-Ready Education Systems

INTEGRATE EDUCATION FOR CLIMATE ACTION INTO EACH SUBSECTOR

Common Directions for Rolling Out Education for Climate Action Across Subsectors

All education subsectors need to teach E4CA. Each education subsector[4]— basic and secondary education, TVET, and higher education—should play its role in teaching climate literacy and green skills (Figure 5). Basic and secondary education target climate literacy as part of general education. TVET provide students with green skills needed for jobs in the low-carbon economy. Higher education provides advanced specialized climate knowledge across various disciplines including engineering, humanities, social science, law and finance.

Figure 5: Overview of Education for Climate Action Across Subsectors

	E4CA in basic and secondary education	E4CA in TVET	E4CA in higher education
Policy objective	Prepare learners for living in climate-resilient societies	Prepare learners for jobs in low-carbon economies	Prepare learners for jobs and life in climate-resilient and low-carbon economies and societies
Learning goal	Transformative climate literacy	Green job skills	Green job skills and transformative climate literacy
Key skills	Transferable and transformative cognitive, social-emotional, and behavioral skills	Climate technology- and task-specific skills; basic environmental knowledge	Discipline-specific knowledge and transformative cognitive, social-emotional, and behavioral skills

E4CA = education for climate action, TVET = technical and vocational education and training.
Source: Author.

---

[4]   The relevance of integrating climate change education also in early childhood education is acknowledged. The focus of the publication is on basic and secondary education, TVET, and higher education.

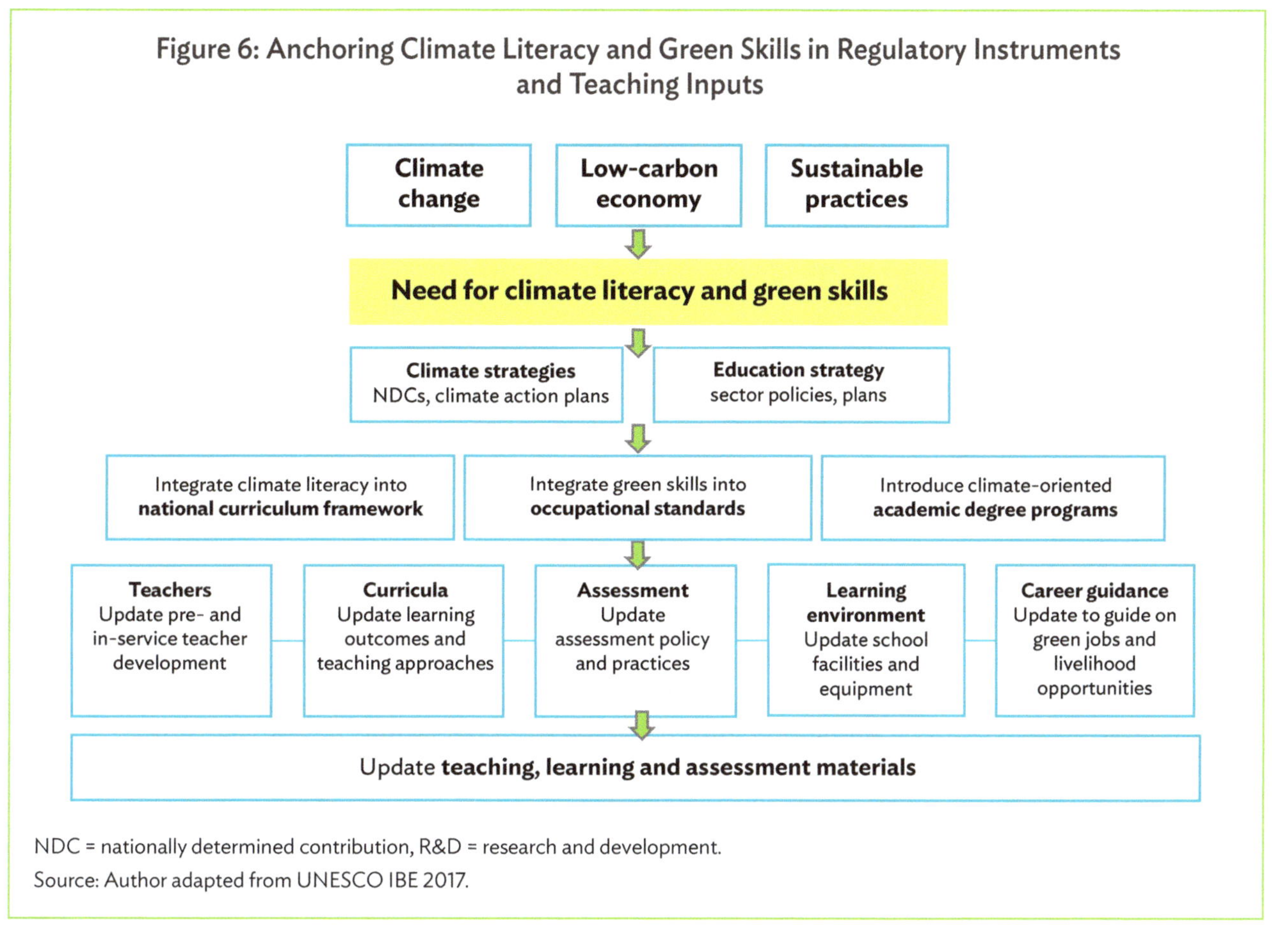

Figure 6: Anchoring Climate Literacy and Green Skills in Regulatory Instruments and Teaching Inputs

NDC = nationally determined contribution, R&D = research and development.
Source: Author adapted from UNESCO IBE 2017.

E4CA needs to be anchored in education regulatory instruments and teaching inputs. All education subsectors need to go through an upgrading and capacity development process to introduce E4CA. Climate competencies for climate literacy (Table 2) and green skills need to be translated into learning goals in national curriculum frameworks or occupational standards in vocational education. Climate learning goals are in turn incorporated into syllabi and curricula. Teacher capacity, pedagogical approaches, assessment methods, teaching and learning material (TLM) and teaching equipment are adjusted accordingly to teach climate literacy and green skills effectively (Figure 6).

The whole-of-school approach (WSA) provides a comprehensive framework to roll out E4CA on the school level. The rollout of E4CA also requires capacity and organizational development on the school level, which may follow WSA. The approach suggests aligning four interrelated components of schools to E4CA: (i) school governance; (ii) school facilities and operations; (iii) teaching and learning, including curriculum, pedagogy, and teachers; and (iv) communication engagement and partnerships (UNESCO 2024a). The UNESCO Green School Quality Standard identifies around 190 actions across the four components that schools could perform to effectively teach E4CA in and outside classrooms (Figure 7) (UNESCO 2024a). A similar guide has been put forward for TVET institutions (UNESCO-UNEVOC 2017). Evidence shows that following the WSA[5] has a statistically significant effect on students' and teachers' motivation, empowerment, and sustainable behavior (Holst, Grund, and Brock 2024).

---

[5] Schools that apply a WSA implement specific actions such as action-oriented learning, applied learning in context of the local community, resource-efficiency monitoring of school operations, critical self-evaluation, recognition for climate engagement, targeted staff development, and climate change as a topic outside of classrooms.

Figure 7: Four Entry Points for Implementing a Whole-of-School Approach

School Governance	Facilities and Operations	Teaching and Learning	Community Engagement
1. Cultivating sustainable practices 2. Ensuring daily sustainable practices 3. Resilience and climate proof governance 4. Establishing a green community	5. Climate education, awareness, and training 6. Developing a climate-friendly infrastructure 7. Ensuring climate resilience and disaster preparedness 8. Promoting green procurement and ethical purchasing	9. Integrating ESD in teaching and learning 10. Fostering meaningful connections beyond the school 11. Hands-on projects and initiatives 12. Leadership and capacity development	13. Building climate resilience in the community 14. School's contribution to community resilience to climate change 15. Local community support for climate change education 16. General community-based climate awareness

ESD = Education for Sustainable Development.
Source: Author adapted from UNESCO 2024a.

The effective rollout of E4CA requires inclusive and gender-responsive approaches. Historically, women and girls have been excluded from many domains that are critical for the low-carbon economy and are disproportionality affected by climate change. For example, the majority of anticipated green jobs are in construction and trade occupations, where women are traditionally underrepresented (IEA 2022b, ADB 2023b). Women are underrepresented in science, technology, education, and mathematics (STEM) and research, which are crucial fields in E4CA (UNDP 2024). Extreme weather events tend to disproportionally affect women's health, for example due to higher sensitivity to the effects of (extreme weather-induced) food insecurity, and poor sanitation during menstruation or pregnancy (Sorensen et al. 2018). Therefore, it is critical to use gender-responsive and inclusive approaches when making education system climate-ready to ensure equitable access to E4CA and the low-carbon and climate-resilient economy for women and girls.

Each education subsector faces unique directions for rolling out E4CA. The anchoring of E4CA in regulatory instruments, the WSA, and inclusive and gender-transformative approaches are shared themes for rolling out E4CA across education subsectors. At the same time each education subsector faces unique challenges that require targeted strategies for rolling out climate literacy in basic and secondary education, green skills in TVET, and advanced knowledge and skills in higher education. These strategies are outlined in the remainder of the section.

CALL TO ACTION—MONITORING AND EVALUATION

Ensure that in all new education projects, measures and metrics are incorporated that track and monitor E4CA and related capacity and skills development of beneficiaries—including girls, women, and disadvantaged youth.

Education for Climate Action in Basic and Secondary Education—Preparing the Youth for Climate-Resilient Societies

National competencies frameworks need to include learning outcomes for transformative climate literacy. Basic and secondary education aims to provide children and youth with general knowledge and life skills which include—in the age of climate change and low-carbon society—also transformative climate literacy. There is an emerging consensus that these learning outcomes are multifaceted, and encompass three key learning objectives (i) the knowledge to understand climate change as it interacts with socioeconomic and socio-ecological systems, (ii) the ability to act upon climate risks in ways that mitigate against further emissions and adapt to the impacts of climate change, and (iii) the ability to negotiate ecological and economic imperatives while centering equity and justice (Bianchi, Pisiotis, and Cabrera Giraldez 2022; UNESCO 2024b). Transformative climate literacy competencies need to be integrated into curricula as new learning outcomes and/or updated existing learning outcomes. Each country can set out its own competencies framework for transformative climate literacy that meets context-specific climate change challenges, incorporate a localized understanding of sustainable living and solutions, and fit the structure of the national education system (Reimers 2021).

Transformative learning is key to E4CA. The objective of transformative climate literacy is to change how individuals understand and interact with the environment informing actions with guiding principles based on sustainability, climate resilience, circularity, biodiversity, collective agency, and equality. This shift requires transformative learning that transforms learners' perspectives, behaviors, and decision-making frameworks, and aligns them with these principles. It therefore includes, besides climate science knowledge, also higher cognition skills such as socio-emotional competencies required for the agency and capacity to take responsibility for climate and collaborate with others for climate action (Reimers 2021). Climate education aims to replace outdated knowledge and competencies that are geared toward the skill needs of carbon-intensive economies that enable unsustainable resource use, environmental degradation, and systemic inequality.

Climate education teaches a breadth of competencies and skills that promote transformative climate literacy. Given the transformative learning needs, learners must develop a range of subject specific, transferable, and transformative skills. This includes subject-specific skills such as interpreting climate-related data and risks, transferable skills such as future-thinking and problem solving, and transformative skills including coalition building, civic engagement, and solidarity (Table 2). Importantly, transformative skills enable learners to promote equity and a just transition through the ability to understand the socioeconomic dimensions of climate change and its impact on different groups, to value multiple perspectives, and recognize how inequality creates differentiated access to adaptive capacities. Therefore, transformative climate literacy provides learners with the competencies for climate actions that are not only scientifically sound, but also just and culturally appropriate in line with local belief systems and cultural identity.

Table 2: Skills Needed for Transformative Climate Literacy

Competencies	Climate Literacy Skills
Subject-specific competencies	• Analyzing climate-related information (e.g., emissions data, environmental impacts) and assessing climate risks and opportunities (climate data analysis) • Understanding how products are made, used, and disposed of, and evaluate their environmental impact throughout their life cycle (life cycle analysis) • Making informed decisions that balance environmental, social, and economic factors in the long-term (sustainability-driven decision-making) • Understanding technologies, approaches, and behaviors that reduce energy consumption and improve resource efficiency and climate risks (energy efficiency)
Transferable competencies	• Understanding the interconnectedness of social, economic, and ecological systems in the context of climate change (systems thinking) • Anticipating and planning for future climate scenarios (future-thinking) • Analyzing and evaluating climate data and policies, and assessing their impacts on communities, and vulnerable populations, including women and girls (critical thinking) • Developing innovative solutions to complex climate challenges at local and global scales (problem-solving) • Assessing the scientific accuracy of climate data and methods (scientific rigor)
Transformative competencies	• Recognizing one's role within a larger community of individuals and ecosystems (climate identity) • Reflecting on and integrating one's own and other's values and viewpoints (valuing multiple perspectives) • Interacting with diverse groups to co-develop climate solutions and effectively guide and implement transformative actions (collective action) • Bringing together diverse groups of people (e.g., Indigenous Peoples) and to engage and facilitate climate-related dialogue (coalition-building)

Author: Based on Brundiers et al., 2021; Kwauk and Casey, 2022.

Climate education can be integrated into existing curricula across all subjects. Climate education can be integrated into curricula in all subjects including the sciences, social sciences, the humanities, and arts, while it can be taught by subject as well as by interdisciplinary methods (Table 3). It is important to take an integrative approach to avoid overload of curricula. Climate learning outcomes need to be integrated into subject domains that encompass environmental (climate science and ecosystems and biodiversity), social (resilience-building and climate justice) and economic topics (low-carbon economies and sustainable lifestyles) (UNESCO 2024b). This also includes gender equality messaging with emphasis on equality and climate justice. One example of this is the ADB-supported Senior Secondary Education Improvement Project in Solomon Islands, which supports the holistic integration of E4CA into curricula across most subjects in senior secondary education (Case Study 1). The UNESCO (2024b) Greening Curriculum Guidance outlines exemplary climate competencies across diverse subject areas including arts, music, and physical education—subject areas that are particularly suited for socio-emotional learning, play-based learning, and other embodied pedagogies that integrate alternative ways of knowing (Bentz and O'Brien 2019, Eusterbrock 2024).

STEM education requires climate reorientation. As STEM education is a crucial part of E4CA, it needs to be reoriented toward climate literacy. The Organisation for Economic Co-operation and Development's (OECD) Programme for International Student Assessment (PISA) 2025 science framework, for example, underlines that science education needs to teach scientific literacy as a core skill, while also contextualizing it with the socio-ecological challenges of climate change (OECD 2024). Science education need foster the ability to (i) understand the complexity of earth's systems and its interactions with human systems; (ii) critically evaluate different information sources about socio-environmental challenges and solutions; and (iii) make informed choices for addressing complex socio-ecological, which will require interdisciplinary teaching approaches. The focus of climate science education extends over science knowledge only and focuses on learner's self-efficacy and constructive engagement to real-life climate problems.

Gender equality must be addressed in STEM education. Globally only around 30% of female students choose to study STEM in higher education. In Southeast Asia, the share of women graduating in STEM varies from about 17% in Cambodia to 37% in Indonesia (Joffre and Serafica 2023). The narrative around gender stereotypes must change to encourage girls in basic and secondary education to pursue STEM subjects and careers. An example of orienting STEM to E4CA is the ADB-supported Improving the Science, Technology, Engineering, and Mathematics Secondary Education Project in Tajikistan[6] which, among others, supports the development of gender-responsive climate change modules for adoption in teaching and learning material and in the pilot preservice teacher education program.

Students learn about climate science at the Oudong High School in Veang Chas, Cambodia.

---

6   ADB. 2023. Tajikistan: Improving the Science, Technology, Engineering, and Mathematics Secondary Education Project.

Table 3: Examples of Education for Climate Action Learning Outcomes in Different Subjects

Subjects	Learning Outcomes
Science	• **Cognitive:** Understand how solar ovens obtain energy. • **Socio-emotional:** Discuss solar ovens with a peer in relation to scientific (e.g., heat, reflectivity), socio-economic (e.g., lack of access to other fuel types), and environmental (e.g., less fossil fuel emissions) considerations. • **Action-oriented:** Build a solar oven and cook a food item.
English	• **Cognitive:** Research the benefits and challenges of a "just transition." • **Socio-emotional:** Keep a journal related to "just transitions" that includes emotions felt, as well as an examination of prejudices, biases, values, and thoughts. Write a reflection about your findings related to "just transitions" from the community interviews. • **Action-oriented:** Record a podcast that interviews community members about their perspectives on a "just transition."
Social Studies	• **Cognitive:** Investigate the cultural-psycho-socio-economic considerations of renewable energy, including in relation to Indigenous priorities. • **Socio-emotional:** Write a letter to a government official about renewable energy actions you would like taken and why these actions are important to you. • **Action-oriented:** Research renewable energy actions taken within your city/state and then create a renewable energy action plan for your city.
Art	• **Cognitive:** Investigate the role of artists in awareness-raising and taking climate change action within society. • **Socio-emotional:** Discuss how a work of art is related to climate change action, as well as how artistic techniques are used to communicate about the issue represented in the artwork (e.g., use of color to portray emotion). • **Action-oriented:** Create a work of art to raise awareness about climate change actions, including in relation to the climate justice movement and Indigenous peoples.

Source: Based on MECCE and NAAEE 2022.

Pedagogy and assessments need to be tailored to achieve E4CA's intended transformative learning outcomes. The way E4CA is taught matters for learning outcomes. Key pedagogical approaches that are particularly effective at fostering climate action-oriented competencies include place-based and action-oriented education and experimental learning (Table 4) (Monroe et al. 2019; Ardoin, Bowers and Gaillard 2020; Rousell and Cutter-Mackenzie-Knowles 2020). These pedagogical approaches enable contextualized, empowering, and reflexive learning experiences that are key to achieving transformative climate literacy and should be combined with gender-responsive and socially inclusive pedagogies. Creating real-life spaces for learning and climate action is critical to empowering youth to overcome feelings of apathy and powerlessness. Given the need for tailored pedagogical approaches for E4CA, institutes for pedagogy are developing guides for teachers on green pedagogy (UCAEP 2018). In addition, formative and summative assessment approaches need to be adjusted corresponding to intended climate competencies and may include assessment methods such as reflective writing, scenario test or concept mapping (European Commission 2022).

Table 4: Key Pedagogies for Education for Climate Action

Relevant Pedagogy	Description
Place-based	Teaching draws on locally and personally relevant dimensions of climate change, centering learning in the lived experiences of climate and environmental change by learners themselves and their local communities.
Community-engaged	Teaching extends beyond the classroom to create opportunities for collaboration and partnership with scientists, community organizations, parents, and/or businesses to facilitate real-world learning and climate problem solving.
Action-oriented	Teaching integrates local-scale action, like climate action projects at school or in the community, which students design through inquiry-based learning and implement in partnership with scientists and/or community members.
Intersectional	Teaching addresses the complex ways in which climate change intersects with social, economic, and environmental injustice and builds understanding of how climate impacts are experienced unequally based on factors like gender, race, socioeconomic status, geography, etc.

Source: Author.

The introduction of E4CA and teacher professional development requires support through universities. Teachers require support to put new learning outcomes, pedagogies, and assessment methods into practice. Support can include capacity development (pre- and in-service teacher education), updated and locally relevant teaching, learning and assessment materials, and access to supportive peer networks. In addition, school management needs to lead by example and support its teachers by facilitating school partnerships with communities, universities, and local businesses to help extend learning beyond traditional classroom boundaries. Universities such as faculties for pedagogy and teacher training institutions possess the institutional resources to directly support schools and teachers in introducing E4CA and need to be key partners of education authorities when introducing and building capacities for E4CA (Reimers 2021).

CALL TO ACTION—TRANSFORMATIVE CLIMATE LITERACY

Ensure that in all new basic and secondary education projects, basic climate literacy for all and transformative and gender-responsive E4CA are incorporated in the majority of curricula.

Education for Climate Action in the Skills and TVET Sector—Prepare Youth and Workers for Jobs in the Low-Carbon Economy

Green skills are needed across all industries. The low-carbon economy changes skills requirements across the labor market. Various climate solutions (e.g., electric vehicles, heat pumps, battery storage, green hydrogen, solar photovoltaic, etc.) as well as environmental principles (e.g., resource efficiency, circularity, sustainable use, and/or biodiversity) are being adopted across industries, including in the clean energy and e-mobility sectors. Many industries will implement measures to decarbonize and lower pollution levels often required by environmental regulation. For example, the steel industry is not conventionally regarded as a green industry. Yet in a drive for decarbonization and circularity, the industry has identified environmental awareness, energy efficiency, and waste management as emerging skill areas, and sustainability manager, recycling specialist, and environmental engineers as emerging job profiles in the industry (Schröder 2023).

Green jobs[7] and green skills are diverse and require targeted updates to qualifications. As the low-carbon and climate-resilient economy entails a wide range of climate technologies and environmental processes, green skills needs are diverse. They include (i) sector-specific green jobs often associated with specific green technologies such as solar panel technician or hydrogen engineer, (ii) sector-agnostic green jobs such as sustainability manager, and (iii) conventional jobs that are complemented with green skills such as sustainable procurement specialist and climate aware health workers. These different green jobs require different green skills that can include technical skills such as technology-specific skills and specialist environmental knowledge, and core skills such as basic environmental knowledge (Table 5).

---

[7]   Green jobs are decent jobs that contribute to protecting and restoring the environment and addressing climate change. Green jobs can be found in both the production of green products and services, such as renewable energy, and in environmentally friendly processes, such as recycling. Green jobs help improve energy and raw material efficiency, limit greenhouse gas emissions, minimize waste and pollution, protect and restore ecosystems, and support adaptation to the impacts of climate change.

Table 5: List of Different Green Job Skills

Green Job Skills	Examples
Technology-specific green skills	Troubleshooting of a solar panel
	Repair and maintenance of electric vehicle
Specialist environmental knowledge, often sector-agnostic	Environment, social, and governance compliance
	Energy-efficiency auditing
Environmental knowledge that complements an existing competency	Sustainable procurement
	Facility energy use management
Basic environmental knowledge as a core skill	Minimization of environmental damage
	Environmental compliance

Source: Author.

It requires targeted updates to qualifications. Given the different green skills needs across occupations, education planners need to have a clear understanding of industry needs when updating qualifications and occupational standards. Some jobs change completely requiring new qualifications, others change moderately requiring some changes to existing qualification or the rollout of short-term professional up-skilling trainings, while yet again other jobs may not change, but require basic knowledge on sustainability (ADB 2023b). In dialog with business industry, new skill requirements need to be identified and an qualification updating approach agreed upon (Figure 8).

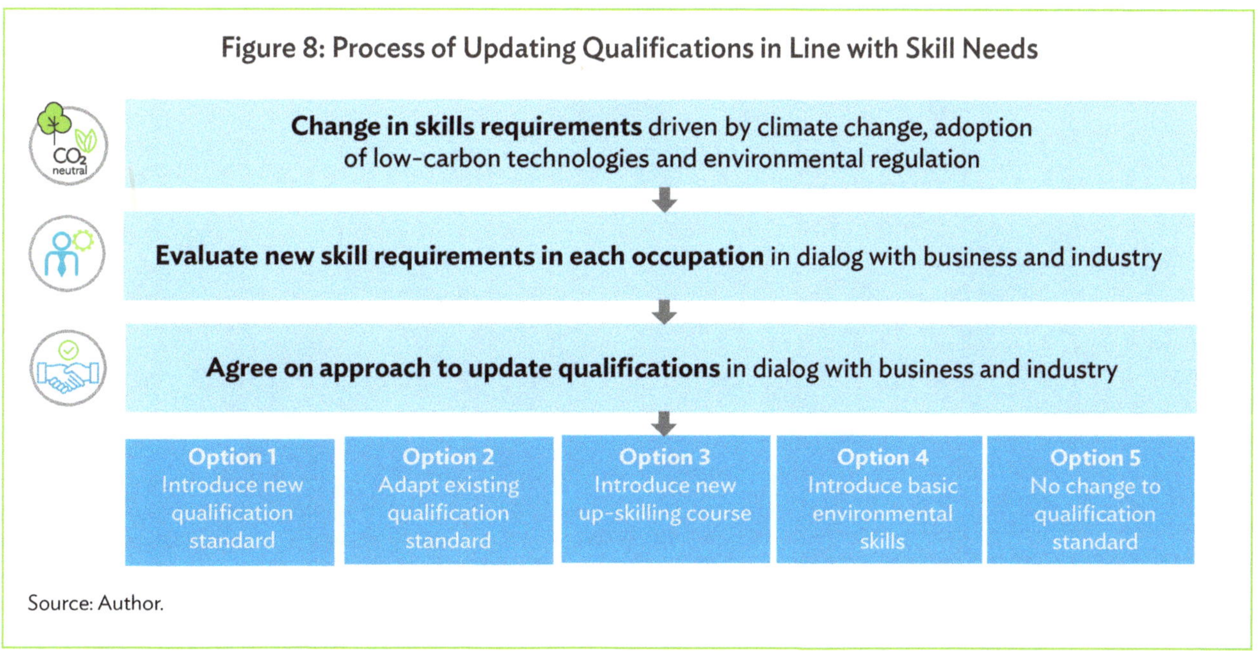

Source: Author.

Basic environmental knowledge is a core skill across all jobs. All jobs in the economy need to be performed in a way that minimize environmental damage and maximizes resource-efficiency. Consequently, there is a need for workers who possess climate related knowledge, skills, attitudes, and competencies that enable them to conduct a job and work tasks environmentally friendly, which may be referred to as climate-smart workers (Climate and Industry Research Team 2023). The International Labour Organization (ILO) Global framework on core skills for life and work in the 21st century regards universal sustainability skills as a core skill (ILO 2021). Thus, basic environmental core skills need to be integrated across all occupational standards. An example of this is the standard core curriculum for environmental protection and sustainability in the German TVET system applicable to all vocational occupations (Box 1).

Soft skills, digital skills, and basic climate literacy are part of green qualifications. Soft skills, such as communication, teamwork, and valuing indigenous knowledge remain important skills in the labor market and are therefore part of green qualifications. Digital skills are also in demand in green industries such as remote sensing, smart metering, and irrigation automation. In addition, TVET systems that include general education in vocational programs (e.g., citizenship education) and hence have space in the curriculum, can additionally integrate climate literacy competencies into curricula as a life skill (see section on basic and secondary education).

A student develops a model for an automated greenhouse at the SSI College Iberia in Kutaisi, Georgia.

Green entrepreneurship skills can support climate-oriented self-employment. In the context of increasing youth unemployment across Asia and the Pacific, entrepreneurship training is increasingly important to equip TVET students with the skills and attitude needed for self-employment. Green entrepreneurship skills generally require the same competencies as regular entrepreneurship such as the ability to identify opportunities, develop ideas, and think along business principles, as well as personality traits such as resilience, self-efficacy, and disruptive thinking (Bacigalupo et al. 2016). To orient entrepreneurship skills toward climate solutions they need to be supplemented by climate literacy competencies such as valuing and understanding sustainability, visioning sustainable futures, and understanding environmental regulation and technology paradigms (Growing Green 2024).

Green qualification programs may incorporate different green skills. Given the diversity in green skills and the importance of soft skills, long-term TVET programs and short-term skilling programs may incorporate different combinations of skills depending on the nature of each occupation. These skills can include climate technology-specific skills, basic environmental knowledge, soft skills, digital skills, entrepreneurship skills and climate literacy competencies. One example of this is the ADB-supported Assam Skills University project that supports the development of various training programs that include a common climate change module, soft skills as well as job-specific green skills (Case Study 2).

**Box 1: Core Curriculum for Environmental Protection and Sustainability
Across All Occupational Qualifications in Germany**

The German technical and vocational education and training system has introduced a standard core curriculum for environmental protection and sustainability applicable to all new and updated occupational qualifications. The core curriculum outlines six sustainability core competencies statements that teachers need to contextualize to the given occupation they are teaching:

- Identify possibilities to avoid damage to the environment and society caused by economic activities in own's one work area and develop them further.
- Use materials and energy under consideration of economic, environmental and social dimensions, in context of work process in production and services.
- Comply with regulations for environmental protection.
- Avoid waste and ensure that substances and materials undergo environmentally friendly recycling or disposal.
- Develop suggestions for sustainable action in one's own work area.
- Collaborate in line with economic, ecological and social sustainable development, and target-group specific communication.

Source: Author based on Kastrup and Kuhlmeier 2023.

Utilities and national priority industries provide key entry points for rolling out green skills training. Green skills needs are gradually emerging across economies. In some industries, green skills needs have already accelerated—driven by fast technology developments, dynamic changes in global value chains, environmental regulations, and policy incentives. For example, in the energy and water sector, utilities are often first movers in installing hydropower plants, wind-energy parks, and sustainable water treatment technologies. In the transport sector, bus operators replace fossil fuel-based vehicles with electric buses and install charging stations at a fleet-level scale. In the textile and garment industry, manufacturers, driven by demands in global value chains, introduce resource-efficient manufacturing processes such as waterless wet processing that reduces chemical waste (Case Study 3). The health sector faces urgent green skills needs, given the immense impact of climate change and extreme weather events on public health. Health workers require climate-specific competencies to identify climate-related health risks, provide climate-informed health advice and care services, and engage in preventive and adaptive actions. Promising entry points need to be strategically identified and leveraged to accelerate the rollout of green qualification programs where they are needed (European Training Foundation 2024).

Green job forecasting and skills gap assessments in dialogue with industry are key to identifying needed changes to qualifications. To systematically account for green skills needs in the economy, workforce planners should anticipate emerging skills requirements to allow for timely updates of qualifications and avoid skills shortages. The updating of qualifications requires skills forecasting. Methodologies for skills forecasting can combine qualitative and quantitative methods such as scenario-based skills forecasting, industry surveys, Delphi methods, macroeconomic modeling, and big data analysis of online job vacancy portals where available (ETF, European Centre for the Development of Vocational Training, and the International Labour Office 2016). Building capacities of workforce planners, labor market institutes, and statistical offices in these methods is required to ensure the accuracy of skills forecasting (CEDEFOP and OECD 2022). Countries that can effectively forecast skill needs can reduce labor shortages before they occur. In addition, gender-sensitive forecasting can provide critical insights into gender disparities in the access to green jobs. Green jobs offer significant potential to empower women, who are often underrepresented in emerging green sectors. To close the gender gap in green jobs, targeted skills training and equal access to reskilling are essential.

Two female staff conduct operations and maintenance works at the Vena Energy Solar Farm on Lombok sland, Indonesia.

Mandated institutions to organize industry dialogue and conduct workforce assessments are key.
Labor market analyses, data science, and structured industry dialogue are key instruments in a labor
market-oriented TVET and skills system. It requires institutions that are mandated, funded, and have the
technical expertise to implement them (World Economic Forum 2019). An example of this is the Sector
Council for Green Jobs in India, an initiative led by the Ministry of Skills Development and Entrepreneurship
(SCGJ 2024.). The council has a permanently established organizational structure, a public–private
and cross-ministerial governing body, and its mandate includes conducting occupational mapping and
skills gap analyses (see for example for the hydrogen sector Chaturvedi, Goyal, and Agarwal 2024),
developing qualification standards for 16 green industries across the energy, water, waste, construction,
and transport sector.

**Investing into life-long learning and skills system enables a just transition and workforce upskilling
across industries.** The phasing out of emission-intensive technologies and industries such as coal and the
internal combustion engine, are regarded as a necessity to meet climate targets set out in NDCs. The skillset
of many workers in emission-intensive industries will, at least partially, become redundant in the low-carbon
economy. Consequently, re- and upskilling support will be a critical component in assisting workers to
transition into new quality jobs and livelihoods, ensuring a just transition in which no one is left behind.
In addition, life-learning learning is also critical to support industries and micro, small, and medium-sized
enterprises in upskilling their workers when adopting climate technologies and solutions. Governments need
to invest into strong skills systems that provide easy access to lifelong learning, informs adult learners about
career and training opportunities, provides incentives for upskilling, and can effectively identify, target, and
support those workers and companies that are affected by the transition to a low-carbon economy.

TVET delivery remains centered around practical skills and work-based learning. Learner-centered
and work-based learning, such as on-the-job training, remain key teaching and learning methods for green
skills development in TVET ensuring the industry relevance and employability of students. Thus, the
introduction of effective green skills training requires strengthening of learner-centered practical training
and the cooperation with business and industry for the provision of work-based learning. TVET institutes
require support and capacity building including teacher capacity development, updated guidelines for
lesson planning, teaching and learning material, and climate technology-specific teaching equipment
among others. In addition, the cooperation with business and industry needs to be strengthened through
the promotion of industry partnerships, capacity building of in-company trainers, and policies that provide
an enabling environment and formal structure for work-based learning.

CALL TO ACTION—GREEN SKILLS

Ensure that in all new technical and vocational education and training, skills, and
workforce development projects, green and climate resilience skills are introduced
in occupational qualifications, linked to key economic sectors in a country's priority
industries, and aligned with NDCs, national climate action plans, and economic
development plans.

Education for Climate Action in Higher Education— Prepare Learners for Jobs in a Low-Carbon Economy and for Life in Climate-Resilient Societies

Higher education provides green skills and knowledge across various sectors. Higher education can support the transition to a low-carbon and climate-resilient economy and society across various disciplines such as engineering, humanities, agriculture, and business. Universities play a critical role in offering degree programs that provide skills for highly skilled green jobs in the service sector such as climate finance, environmental policy, teacher education, sustainable urban planning, environmental law, and sustainable business management, as well as the technology sector such as hydrogen engineering and climate science. These green service jobs tend to be critical for, among others, business, regulation, policy, and project development in the low-carbon and climate-resilient economy.

Universities have several options to incorporate E4CA into their degree programs. Universities can leverage their flexibility in academic structure to integrate E4CA into degree programs in various ways by either integrating it into existing teaching or creating new teaching offers (Figure 8). Leveraging

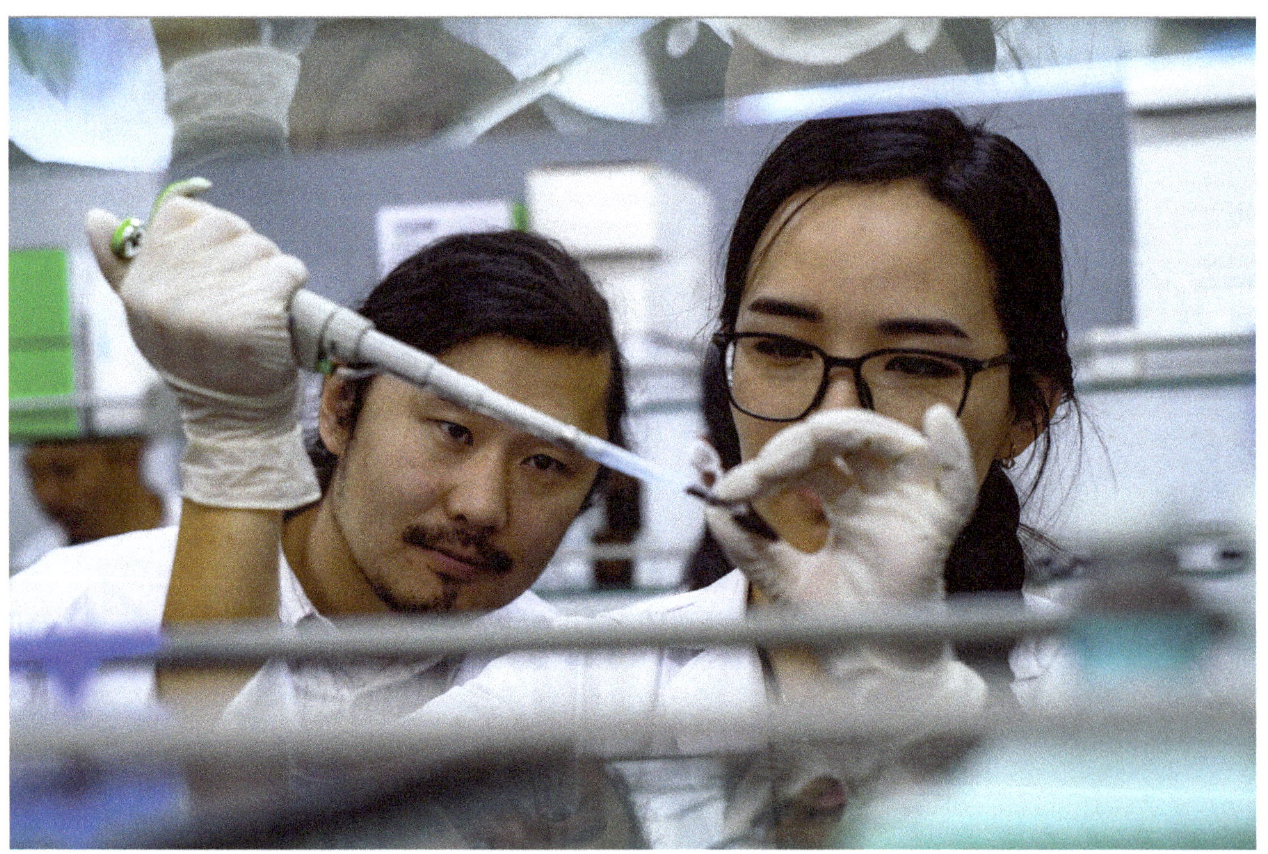

Students conduct an experiment in a genetic engineering laboratory at the National University of Mongolia.

existing structures universities may integrate climate-related content into existing courses, either with a narrow approach integrating specific climate modules into specific courses ("piggybacking") or with a broader approach interweaving climate learning goals into existing curriculum across multiple courses ("mainstreaming"). Another option is to develop new structures either with a narrow approach creating specialized degree programs that focus specifically on climate issues ("specializing") or a broader focus on establishing inter- and trans- disciplinary programs that connect E4CA with broader themes across different fields ("connecting") (Figure 9). The choice of approach depends on the specific training needs. Piggybacking and specializing can be implemented relatively quickly with immediate benefits, whereas mainstreaming and connecting require more extensive and coordinated efforts with the benefit of providing a more integrated and systematic approach to E4CA. Universities can employ a combination of these strategies as they are not mutually exclusive.

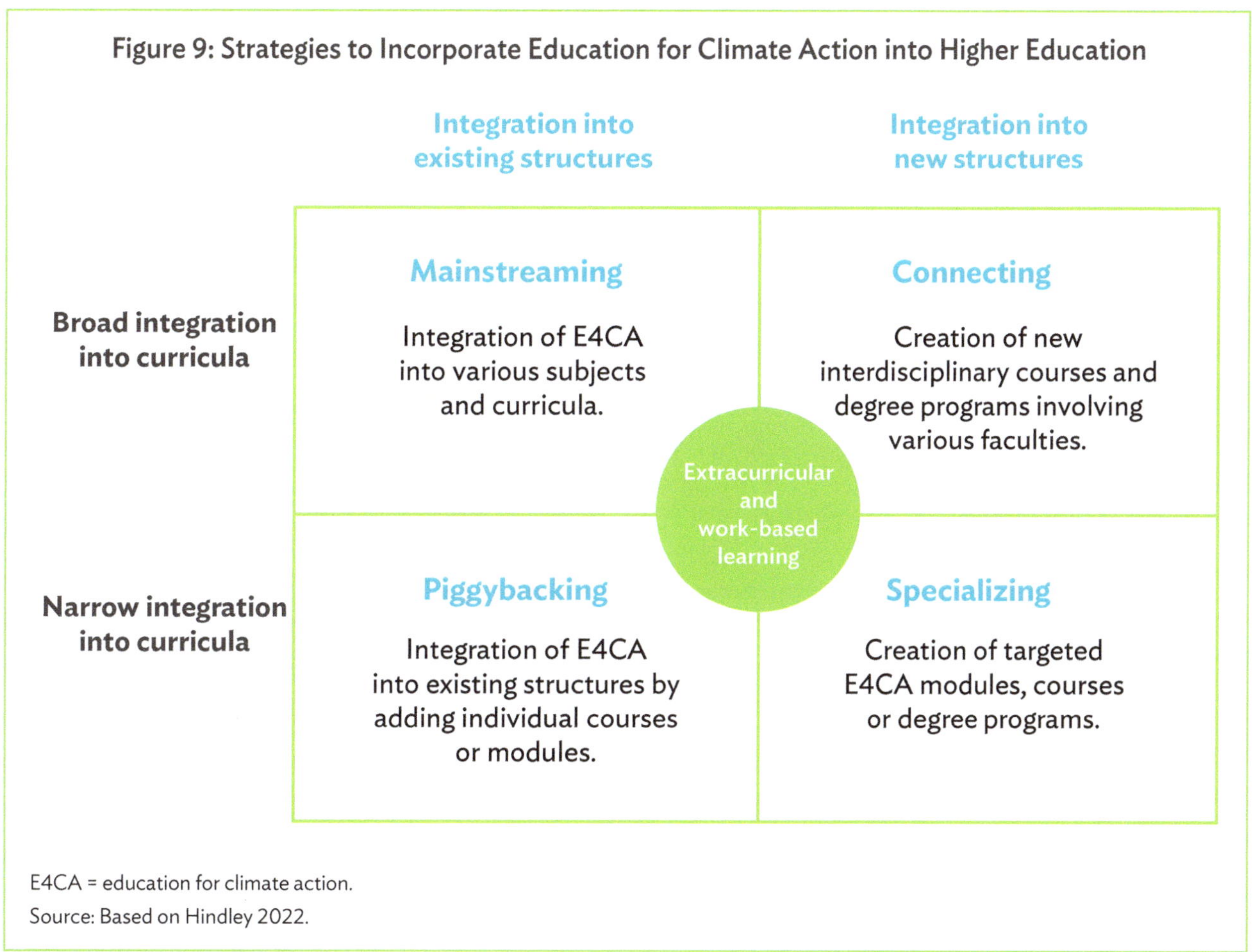

Figure 9: Strategies to Incorporate Education for Climate Action into Higher Education

E4CA = education for climate action.
Source: Based on Hindley 2022.

Besides degree-specific E4CA, higher education can also teach climate literacy as a core skill.
Climate literacy can also be taught as a core skill at higher education institutions targeting similar learning goals as in basic and secondary education (Table 2). An example of climate literacy at the tertiary education level is the introduction of an interdisciplinary collaborative core (ICC) requirement in the undergraduate curriculum at the Nanyang Technological University (NTU) Singapore (NTU n.d.). NTU has introduced seven new core courses including sustainability, healthy living, interdisciplinary inquiry and communication skills, and science. These core courses are mandatory, credited, and scheduled over the first 2 years of study making up between 20%–30% of the total undergraduate curriculum.

Universities need to evolve organizationally to provide interdisciplinary education. Interdisciplinary programs are needed to provide the skills required to address the most complex challenges of climate change in its interaction with the various domains of human lives. Universities are ideally placed to provide interdisciplinary degree programs as they have expertise in various subjects and teach advanced skills. Interdisciplinary education requires new organizational structures within universities such as interdisciplinary teaching centers that bring together academic staff and experts from diverse fields, such as environmental science, economics, social sciences, law, engineering, public health, education. To create such centers, it is crucial to secure sustained funding, institutionalize the center through a dedicated organizational structure and interdisciplinary mission statement, and support faculty staff to engage in interdisciplinary partnerships. An example of this is the Center of Excellence in Climate Change Studies established at the at Rajarata University of Sri Lanka (Case Study 4).

Universities need to access and advance climate knowledge through partnerships. The transition to a low-carbon economy is intertwined with the introduction of new climate technologies and knowledge. To move rapidly into these new knowledge areas, universities need to partner with industry and other academic institutions to be able to offer climate-oriented and interdisciplinary degree programs and research projects. Partnerships need to be strategic based on a shared mission and complementary know-how, facilities and equipment. An example of this is the upcoming global high-tech climate-smart agriculture university network (GHAN) that aims to connect a global hub and regional hubs of agricultural universities in the world to accelerate the speed of technology adoption in each country and regional context. Another example is the Leiden-Delft-Erasmus universities strategic alliance, an alliance of three universities in the Netherlands that have jointly established nine interdisciplinary centers and several interdisciplinary teaching programs (Leiden-Delft-Erasmus n.d.).

University–community partnerships can leverage university resources for local climate solutions. Universities bring a wealth of resources—knowledge, technology, student expertise—that can be leveraged through university-community partnerships to support community-led adaptation efforts and co-create locally relevant climate solutions. Such partnerships not only foster climate resilience in communities but also offer students place-based learning experiences working on real-world problems. One example is the United States National Oceanic and Atmospheric Association Sea Grant College Program. The program funds activities of a network of university–community partnerships and extension center that support coastal communities to build climate resilience through community-led climate adaptation actions and capacities development of the coastal workforce.

CALL TO ACTION—CLIMATE-ORIENTED ADVANCED DEGREE PROGRAMS

Ensure that in all new higher education projects, degree and post-graduate programs are updated or newly developed to include interdisciplinary climate knowledge and competencies in alignment with NDCs and national climate action plans.

ACTION AREA 2

MAKE SCHOOLS CLIMATE-RESILIENT THROUGH CLIMATE ADAPTATION

Making schools climate-resilient is imperative to ensuring educational continuity and preventing learning loss. Among all regions in the world, Asia and the Pacific has the largest exposure to climate change and has incurred the greatest economic losses caused by disasters triggered by natural hazards between 1980–2022 (Dabla-Norris et al. 2023). In particular, low- and middle-income countries face heightened vulnerability to disasters due to greater exposure to climate-related hazards and weaker resilience. Evidence shows (Table 6) that the rise of extreme weather events such as floods, extreme heat, and storms negatively affects learning outcomes. Extreme weather events degrade learning environment (e.g., loud sound of rain on tin roofs, high temperatures in classrooms) and damage schools leading to school closures and undermine the health and livelihoods of children and households with children resulting in lower readiness for learning and early school leaving (Figure 10). For example, students in India who experience prolonged exposure to storms are more likely to experience educational delays, are less likely to complete higher education, and are less likely to secure salaried employment (Box 2). In addition, climate change tends to affect disadvantaged groups the greatest. In Viet Nam 50% of children from the poorest households have experienced at least one extreme weather event by the age of 15, compared to only 17% of children from better-off households (Young Lives 2022). Overall, climate change is a significant challenge to education operations, learning outcomes, student and teacher safety, equality, and can result in significant losses in public assets.

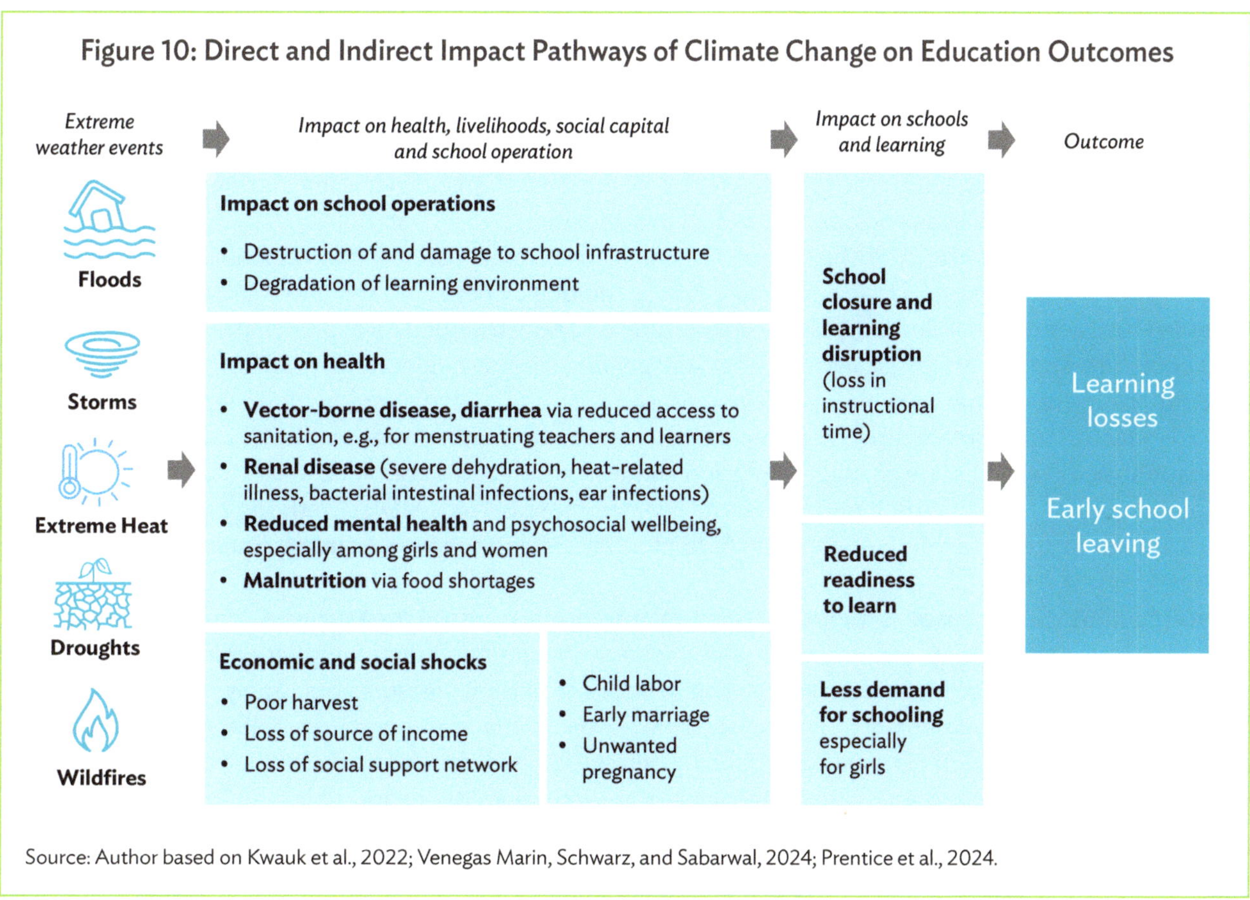

Source: Author based on Kwauk et al., 2022; Venegas Marin, Schwarz, and Sabarwal, 2024; Prentice et al., 2024.

Table 6: Effects of Extreme Weather Events on Learning Outcomes

Study Findings	
Impact of extreme weather events on school operations, infrastructure and learning outcomes	
Flooding	In Pakistan, the floods of 2022 damaged or destroyed at least **17,205 schools** across the country, disrupting the education of approximately **2.6 million children, resulting in $559 million** in damages to education infrastructure (Ministry of Planning Development & Special Initiatives 2022)
Extreme heat	In the Philippines, extreme heat and other calamities caused **32 days of school closures** in the school year 2023–2024 (PIDS 2024)
Impact of extreme weather events on socioeconomic factors and learning outcomes	
Flooding	In India, flooding of the Kosi River in August 2008 caused a 125% increase in **child marriage** and 24% **reduction in secondary school completion** for boys. For girls, the increase in child marriage was 38% and the reduction in secondary school completion was 32% (Khanna and Kochhar 2023)
Flooding	In Viet Nam, reduced household income from flood-related crop failure **directly impacts the amount of time children spend in school**, especially those from poorer households (Nguyen and Pham 2018)
Impact of extreme weather on health and learning outcomes	
Extreme heat	In Brazil students in the hottest municipalities lost about 1% of learning per year. Cumulatively, **losing up to 0.66 to 1.5 years of schooling** by grade 12 due to extreme heat (Schady et al. forthcoming)
Flooding	In India, 6 months after flooding in Kerala in 2018, **60% of students (41% female, 9% male) exhibited symptoms of post-traumatic stress disorder**, impacting their concentration and school attendance (PV and Subudhi 2020)

PIDS = Philippine Institute for Development Studies, UNDP = United Nations Development Programme.
Source: Author.

Climate adaptation measures in schools' infrastructure should be tailored to school-specific climate risk and vulnerabilities. Given the impact of extreme weather events on learning outcomes (Table 6), ensuring that school facilities and operations are climate-resilient to withstand climate impacts and remain operational is as crucial as the education itself. Each school including main and auxiliary buildings, schoolyards, playgrounds, and their surroundings faces unique climate risks that need to be identified, assessed, and addressed following a range of structural and nonstructural measures that enhance climate resilience. Site-specific multi-hazard safety risk and gender-sensitive vulnerability assessments are critical to identify current and future climate as well as disaster risks and design risk reduction measures to make school infrastructure climate resilient. Investing into climate adaptation measures in schools algins the education sector with climate change adaptation and resilience goals as outlined in NDCs, national adaptation plans, as well as the Joint Multilateral Development Banks' (MDB) Methodological Principles for Assessment of Paris Agreement Alignment (ADB et al. 2023a; European Investment Bank 2022).

Climate adaptation measures include structural and nonstructural measures. Adaptation measures help in reducing the impacts of climate risk, allow schools to continue functioning with minimum disruption, and create a culture of safety and resilience. This includes structural measures that apply to all elements of the physical built environment including higher plinth levels to deal with flood risks, walls, passive cooling and vegetation to deal with extreme heat, building orientation and roofs design to reduce risks from tropical cyclones, and rainwater harvesting features to deal with water scarcity. Adaptation solutions should also

Box 2: How Exposure to Tropical Storms and Cyclones Affects Students' Educational Attainment and Labor Market Entry in India

Students' exposure to storms during school years has a significant negative impact on educational outcomes and labor market entry. In terms of educational outcomes, students with an average exposure to storms experience delayed schooling by 7.25% and a decrease in post-secondary attainment of 7.35%. In terms of labor market outcomes, affected students experience an 8% reduction in regular employment, a 4.8% increase in individuals performing domestic duties due to storm exposure, and a decrease of hourly wages by 3.9%.

In the long-term storms lead to an 8% reduction in household income within 10 months post-shock. In addition, the destruction of school infrastructure increases school closures by 7.4% within 2 years. The results underscore the importance of addressing both income support and educational infrastructure in post-disaster recovery. Policies that combine financial aid with incentives for school attendance and improved school resilience, such as building disaster-resistant schools, are essential to mitigating the long-term adverse effects of storms on human capital development.

Source: Pelli and Tschopp 2024.

promote nonstructural interventions that include actions such as developing school disaster preparedness and contingency planning, undertaking disaster safety drills, and improving IT readiness—enhancing the capability for remote teaching. Adaptation activities can also include the broader community such as school disaster preparedness committees involving schools and partners, strengthening school maintenance and operations procedures linked with early warning systems, and including awareness-raising campaigns within wider community on importance of building resilience. All adaptation measures should be gender-responsive to ensure that the needs and vulnerabilities of women and girls are considered. For example, disaster preparedness training should be gender-sensitive, and girls' active participation in school and community-level adaptation efforts such as committees and drill should be promoted. One example is the ADB-supported Disaster Resilience of Schools Project that took a comprehensive approach to climate resilience and disaster risk reduction of schools in Nepal (Case Study 5).

The enabling environment for resilient school infrastructure can be strengthened by improving construction standards and practices. To systemically embed climate adaptation considerations in the school infrastructure design, building codes and school design guidelines can be updated and revised. Close linkages need to be established between the education sector and national institutions involved in updating and setting building construction codes and agencies responsible for providing hazard information. Training and capacity building of planners, civil engineers, architects and operations and maintenance engineers within ministries of education at national and local level, will be critical. While national agencies can provide broad guidance, climate risks are context-specific and need to be identified per school, based on inputs from local experts and population.

Special features where schools are expected to play the role of emergency shelters without resulting in prolonged school closure should be adopted. In developing Asia and the Pacific, schools often end up as emergency shelters. In such context, school infrastructure should be designed to allow the school to play the role of emergency shelter, while minimizing learning loss outcomes. In this context, it is critical that the adaptation-related physical interventions feature the different needs of girls and women and create a safe space for all. In some rural contexts, the design of school emergency shelters would also need to factor other needs of the population such as safe space for hosting livestock.

Adaptive social protection systems can enable continuation of school attendance and learning in the aftermath of climate-related shocks. Diverse range of social protection programs including social transfers such as child grants and social pensions, school feeding programs, and health insurance, are effective at reducing poverty and vulnerability in different contexts and as a result improves school attainment rates and learning outcomes (Table 7). Integrating adaptive features in social protection can help poor and vulnerable households adapt to climate change impacts, without compromising education outcomes. Adaptive features include better climate-risk informed targeting of social protection program beneficiaries, incorporating flexible features in social protection program design to allow horizontal or vertical expansion in the immediate aftermath of disasters, and linked social protection with climate-resilient livelihood interventions. Interministerial and interdepartmental cooperation between social, education, and environmental authorities is key to jointly addressing social vulnerability to climate change such as coordinated education and social sector plans. One example of the effectiveness of social protection programs on learning outcomes has been demonstrated by the Secondary Stipend and Assistance Program in Bangladesh, where a conditional stipend and tuition subsidy program raised women's grade completion by 3.2 years, the secondary completion rate by as much as 5 percentage points, and delayed marriage by 3.2 years (ADB 2021a).

Table 7: Pathways Through Which Social Protection Can Tackle Climate Impacts and Contribute to Improved Learning Outcomes

Climate Risk	Social Protection Instruments	Role in Improving Education Outcomes
Unemployment, change in income status of households due to disasters, and weather-related events	**Social Assistance Programs** (cash transfers, child benefits, school feeding, fee waivers, subsidies)	Financial support to households to continue investments in child's education
	Social Insurance (unemployment insurance, work accidents, livestock insurance etc.)	Prevent income shocks and allow to smooth consumption, without impacting their investments to education
	Social Care (family support services, child care)	Provide support to parents (financial and/or inkind) for care duties, as needed, without impacting them financially, while creating a conducive home environment for child development

Source: Author.

CALL TO ACTION—ADAPTATION

Ensure that all new investments from 2025 in school and education institution infrastructure and operations are adapted to local climate and disasters risks and meet related building standards.

ACTION AREA 3

PROMOTE LOW-CARBON SCHOOLS FOR CLIMATE MITIGATION

The building sector, which includes school facilities, is a central part of most NDCs. In 2023, about 89% of countries in Asia and 75% in the Pacific[8] have NDC building sector measures with a focus on mitigation including for public buildings, which includes public school facilities (PEEB 2023). To align with NDCs and the joint MDB methodological principles for Assessment of Paris Agreement Alignment (ADB et al. 2023a), school infrastructure investments should identify and implement climate mitigation measures that can include (i) energy-efficient building design and equipment; (ii) resource-efficient water, waste, and transport school operations; (iii) installation of decentralized renewable energy; and (iv) energy-efficient school management (Figure 11).

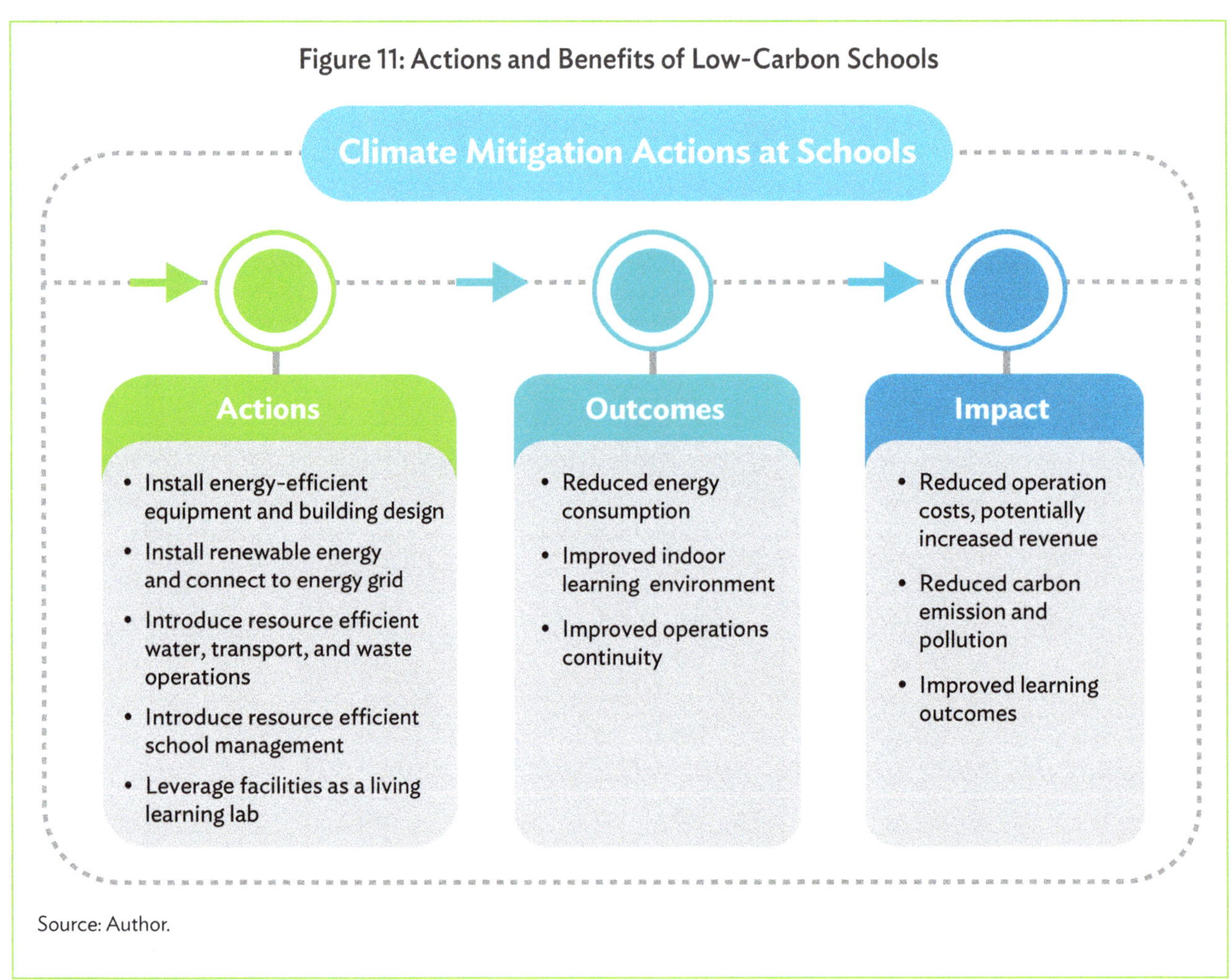

Figure 11: Actions and Benefits of Low-Carbon Schools

Source: Author.

---

[8] References source grouped countries in Asia and the Pacific separately into Asia and Oceania, which includes Australasia.

Climate mitigation measures can result in environmental, economic, health, and educational benefits at schools. Climate mitigation measures at schools provide various benefits: (i) environmental benefits by reducing energy consumption resulting in lower pollution, (ii) economic benefits by reducing operational costs and enhancing building asset value, and (iii) educational benefits by improving comfort and well-being for staff and students resulting in improved learning outcomes (Figure 10). Powering schools with on-site renewable energy has an additional benefit of overcoming unstable power supply providing reliable power for example critical for digital learning. In 2014, in the Philippines, ADB retrofitted 40 government buildings in Metro Manila and 150 government buildings nationwide resulting in 11 gigawatt-hours of energy savings per year, with 4.6-year payback on investment and approximately $2 million of annual cost savings (ADB 2015). In 2022, the refurbishment of schools and kindergarten in Mongolia resulted in around 50% in energy savings per school as well as a decrease in absenteeism of teachers and reduced medical costs (GIZ 2022). Given the various benefits, climate mitigation investments in schools should be included in national infrastructure investment plans for climate mitigation.

Rooftop solar installed at the Islamia University of Bahawalpur in Pakistan providing energy cost savings.

Green building certification provides a structured approach to greening school facilities. Green building certification is not strictly necessary to implement climate mitigation (and adaptation) actions in school facilities. However, they provide a comprehensive framework with clear actions and targets that can be integrated into school development plans and budgets, and that can be audited against objective and transparent criteria. In addition, for newly constructed school buildings green building certifications are required to be considered Paris Agreement aligned according to the Joint MDB Methodological Principles for Assessment of Paris Agreement Alignment (ADB et al. 2023a). Various international as well as national green building certifications can be used such as LEED, BEEAM, EDGE, and the Green Building Council of Sri Lanka certification (Case Study 4). For further information on green building certifications and energy-efficient building designs see ADB's Handbook on Energy Efficiency in Buildings (ADB 2024c).

School retrofitting provides various opportunities for climate mitigation. In the context of school rehabilitation initiatives, schools have various options for implementing and installing individual energy-efficient and renewable energy measures and equipment. The MDBs Common Principles for Climate Mitigation Finance Tracking outline which activities and technologies can generally be regarded as Paris-aligned (ADB et al. 2023b). The principles provide technical guidance for activities that can be found in school operations such as (decentralized) energy generation, waste collection, passenger transport, buildings, public installations and end-use energy efficiency, energy and resource use efficiency and research, development and innovation. Examples of mitigation measures across the areas include solar

A student practices for heavy vehicle driving on a simulator at the Samthang Technical Training Institute, Bhutan.

photovoltaic modules, low-carbon cement, sustainable timber, heat pumps, smart meters, LED lighting, green roofs, green shading, among many others (Table 8). In addition, in vocational teaching subjects that create carbon emissions, the introduction of digital learning can be a mitigation measure. For example, in Bhutan, a digital driving simulator was introduced in the training for heavy vehicle driving and earth mover operation resulting in carbon emission reduction and fuel cost savings (Wangda and Phodrang 2024).

Table 8: Selected Climate Mitigation Actions Across School Operations

School Operations		Possible Climate Mitigation Actions
	Energy-efficient buildings, installations, and end-use energy efficiency	Energy-efficiency equipment. Heating, ventilation and air conditioning (HVAC) such as efficient chillers and boilers, heat pumps, and energy-efficient LED lighting
		Sustainable materials. Use of sustainable timbers, wood, and bamboo and use of low-carbon cement
		Nature-based solutions. Design green spaces, green roofs, green walls and green shading
		Building envelope and materials. Enhance wall and roof insulation, passive energy design, windows with low thermal conductivity/low-emissivity façade glazing
	Renewable energy, energy storage, and grid connection	Install renewable energy technologies on school campus such as rooftop solar photovoltaic for electricity generation and solar thermal for hot water
		Install grid infrastructure (e.g., battery storage, inverters, smart metering)
		Source energy with low life cycle greenhouse gas emissions such as renewable energy, sustainable certified
	Sustainable transport	School fleet. Replace school vehicles with electric vehicles (e.g., e-school buses)
		Promote non-motorized transport or electric personal mobility. Bicycle lanes, campus-based charging station, bike sharing programs
	Circular waste management	Waste management. Introduce waste segregation and recycling practices for paper, plastic, glass and e-waste. Install a composting system
	Water use and conservation	Install water efficient faucets and flushing systems
		Install a rainwater harvesting system for non-potable uses
		Replacement of tanker use for water service delivery with a piped network
	Resource-efficient school management	Install smart electric meters for monitoring
		Introduce green procurement practices
		Introduce digital learning in subjects that require strong resource use or create carbon emissions e.g., vocational education

Source: Author based on SDSN 2022 and ADB et al. 2023b.

Energy-efficiency management must become a core activity of school administration and facility managers in climate-resilient schools. Adopting energy management systems helps ensure energy savings on a day-to-day basis. School-based energy management needs to be driven by school staff including principals, teachers, students, administrators, facility managers and sustainability coordinators (Sustainability Victoria 2016). School staff need to be capacitated and mandated to implement energy management programs. Energy management guidelines for schools are already available such as the *Guidebook for Energy Management in Your School* in India (Bureau of Energy Efficiency – India 2009) and the *Guide to Low Carbon Schools* in Hong Kong, China (Environment Protection Department – Hong Kong 2024). The rollout of energy management programs requires dedicated budget support and capacity development for school administrators, facility managers, and staff. One example is the ADB-supported Guangxi Baise Vocational Education Development Project that supported to transform the Baise University into a sustainable campus (Case Study 6).

Net-zero school facilities can be leveraged for educational purposes through living learning labs.
Climate mitigation measures and energy-efficient school facilities can be leveraged to transform schools into living learning labs turning every aspect of the school's physical and social environment into a climate education opportunity (Robinson, et al. 2021; SDSN, 2022; Preston 2024). Living learning labs provide the opportunity for teachers to teach outside of classroom applying place-based, action-oriented and experimental teaching methods that are key for developing climate literacy. Critical to living learning labs is to use school facilities and equipment as teaching inputs to empower students to actively participate in sustainability initiatives and to take an active role in reducing their carbon footprint. Living learning labs require school facilities and operations with climate-resilience features and teachers with the capacity to integrate them into teaching and learning.

CALL TO ACTION—MITIGATION

Ensure that all new investments from 2025 in school and education institution infrastructure, including newly built schools and renovations, are designed and constructed for resource efficiency, including through renewable energy sources, in line with national building codes and green building standards.

ACTION AREA 4

DRIVE CLIMATE INNOVATION THROUGH INNOVATION HUBS

Universities can support climate solutions innovation and adoption through R&D, incubation, and commercialization support. The successful transition to a low-carbon economy will depend on a country's ability to adopt locally appropriate climate solutions and spur innovation and business development in green industries. National and local innovation systems are a critical infrastructure to support the adoption of climate solutions and development of green industries. Modern universities, in the form of research institutes, incubation and innovation hubs, or science and technology parks, play a central role in innovation systems by providing a range of services. These services include industry-driven research, prototyping, and incubation and commercialization support (Table 9) (ADB 2015). The capacity of universities to spur climate innovation and technology adoption will shape how successful countries will transition to a low-carbon economy.

Innovation hubs need to be developed strategically and tailored to the needs of regional industries and communities. Innovation hubs can support various innovation objectives that vary in the degree of complexity, including promoting green entrepreneurship for self-employment, supporting local industries in the adoption of mature climate technologies, turn research projects into profitable stand-alone businesses, registering green patents for licensing to industries, or supporting industries to move into green industry value chains for industrial development (Clarysse et al. 2005). To ensure a good match with the needs and capacities of local industries, the establishment of innovation hubs should begin with a strategic context assessment that includes defining national or regional needs and assets (e.g., faculty expertise, local industries, sector policies), the development of a business proposition (e.g., priorities, technological focus, strategic partnerships, targeted beneficiaries), service definition, organizational form, funding model, ecosystem vision (e.g., international partnerships, industry networks), and innovation targets (Kalpaka et al. 2020). As innovation sits at the interface of education, R&D and industry and economic development, innovation hub initiatives require interministerial collaboration between line ministries such as the ministry of science and technology, education, industry, and economic development.

Universities need to develop new capabilities to operate innovation hubs. To support climate innovation, universities need to establish knowledge and innovation hubs that go beyond academic-oriented research alone. This transition will require universities to develop new capabilities (Table 9) to support researchers to build entrepreneurship capacities, identify industry-relevant ideas and opportunities, provide access to climate technologies, support incubation as well as commercialization through access to experimentation, funding, and mentorship programs (Table 9) (Sarpong et al. 2023). Additionally, innovation hubs can strengthen local capacity through market and barrier research, organizing knowledge sharing events, and providing specialized education programs that improve the capacity of stakeholders in innovation ecosystem such as innovation policy, environmental law, and climate finance. One example is the ADB supported Promoting Research and Innovation Through Modern and Efficient Science and Technology Parks Project that supports four universities in Indonesia to set up and operate science and technology parks (Case Study 7). Innovation hubs are not limited to universities and can also be operated by skills training hubs.

Table 9: Key Capabilities of Research, Knowledge, and Innovation Hubs

Service Area	Innovation Services/ University Capability	Academic University	Knowledge Hub	Innovation Hub
Academic	Academic research	X	X	X
Applied research and development (R&D)	Green industry-focused research		X	X
	Building a funding pipeline		X	X
	R&D facilities		X	X
	Industry and international engagement		X	X
Innovation system development	Executive education for decision-makers		X	X
	Market barriers and drivers research		X	X
	Knowledge sharing events		X	X
Incubation	Green entrepreneurship training for researchers			X
	Advanced fabrication labs			X
	Entrepreneur mentorship program			X
Commercialization	Access to venture capital			X
	Business and intellectual property advice			X

Source: Author based on United Nations Conference on Trade and Development 2019, Schot and Steinmueller 2018.

Innovation hubs need to be oriented toward climate challenges and opportunities of local industries and communities. To support the transition to a low-carbon and climate-resilient economy, universities need to orient innovation activities toward climate challenges and solutions by for example procuring climate-related equipment, setting up challenge-based research funds focused on climate challenges (Table 10), offering green entrepreneurship training, build partnerships with green industries, and mobilize funding for climate technology grants.

Table 10: Examples of Climate-Oriented Innovations

Innovation Type	Examples of Climate-Oriented Innovations
Product innovation	Development of inexpensive, durable, recyclable, biodegradable, locally sourced building materials
	Turning industrial-scale organic waste material into new valuable products such as fertilizer or animal feed
	Development of an app for remote monitoring and controlling irrigation
Process innovation	Adoption of pollution control and pollution treatment technologies
	Adoption of waste prevention and waste management in local industries
Business innovation	Development of labels informing customer choices and improving supply chain outcomes
	Sustainable supply chain re-design toward local producers, eco-villages

Source: Author based on United Nations Conference on Trade and Development 2019.

Innovation hubs require new capacities and facilities. The establishment of innovation labs at universities often entails the introduction of completely new activities, requiring capacity development at universities. On the infrastructure level, they require collaboration spaces, fabrication labs, and digital and technology equipment. On the human capacity level, innovation labs require a dedicated management and support team with expertise in industry trends, product development, intellectual property, business and partnership development and gender-sensitive innovation services (ADB 2021b). On the institutional level, new procedures for grant administration, revenue-based funding model, professional communication and outreach, monitoring of innovation key performance indicators, and partnership management. At the governance level, they require a dedicated advisory board including members from industry and financial sectors and an ethics committee.

CALL TO ACTION—CLIMATE INNOVATION

Ensure that in all new projects that support research, innovation, and start up incubation, the services of research and development and innovations hubs are climate-oriented, in alignment with NDCs, national climate action plans, and other national economic and innovation strategies.

This climate change and education playbook aims to outline how education policymakers and project developers can make education systems climate ready. This section summarizes the way forward by outlining the (i) key targets of climate ready education systems, (ii) guiding action principles to overcome complex challenges, (iii) and eight calls to action, including actionable strategies, to consider when making education systems climate ready.

Four key targets of climate ready education systems. Future investments need to build climate ready education systems that:

(i) teach transformative climate literacy and green skills across education subsectors including basic and secondary education, TVET and higher education,

(ii) operate climate-resilient school facilities and operations that have been adapted to climate risks,

(iii) operate sustainable and low-carbon school facilities and operations that minimize carbon emission and pollution, and

(iv) promote climate-oriented R&D, entrepreneurship, and incubation that foster innovations in climate technologies and solutions.

Guiding principles to inform action. Future investments in education need to be informed by six guiding action principles to address the complexity and challenges of climate change collaboratively, based on evidence and make cooperation and digital technologies part of solutions in the education sector.

(i) **Mobilize (innovative) finance.** Building climate ready education systems requires investments in human and institutional capacities and school infrastructure. Innovative finance such as Innovative Finance Facility for Climate in Asia and the Pacific Financing Partnership Facility (IF-CAP) and green and blue bonds provide crucial climate finance. Existing climate finance mechanisms need to expand their focus to fund the development climate-ready education system in acknowledgment of the education sector's transformative impact on youth, children, women, workers, communities, industries and innovation systems as well as its critical role in the just transition and enabling actions to achieve various SDGs.

(ii) **Cooperate with business and industry.** Business and industry can be key partners in facilitating the transition to climate ready education system. For example, industries are key partners for introducing and updating green qualification standards; in setting up industry-focused, challenge-based research grants; for running mentorship programs for green entrepreneurs; and, in re- and up-skilling staff in MSMEs that are adopting climate technologies or industries that are phasing out.

(iii) **Develop intersectoral solutions.** Intersectoral approaches can address the complex problems of climate change. For example, coordinating with social protection line ministries to target vulnerable households with school children can improve children's learning outcomes; education line ministries need to closely coordinate with line ministries for the environment, energy, and transport when updating NDCs; and, education planners can cooperate with sector associations, such as utilities in the energy sector to develop tailored workforce development training programs.

(iv) **Promote gender-responsive actions.** All actions must be inclusive and gender-transformative in outcomes, for example, by ensuring gender-responsive school infrastructure designs and learning material and promoting female participation and access to STEM education and green jobs.

(v) **Leverage digital technologies.** Digital technologies can be leveraged to support the transition to climate ready education systems. For example, online learning platform at schools can enhance climate resilience of education delivery by enabling learning continuity during school closures; promoting digital skills can support green jobs such as smart metering and irrigation automation; and digital learning enables quality distance learning reducing travel related carbon-emissions of learners.

(vi) **Conduct high quality analytics.** Analytics and assessments are critical to fuel new nationally appropriate and tailored investments in E4CA and resilience. Key analytics include green labor market forecast, energy audits and climate risk assessment of schools, and strategic green industry and value chain assessments.

Eight calls to action. Strategic actions are needed to make education systems climate ready and introduce E4CA, climate resilience and innovation in a transformative way. These actions are outlined in the eight calls to actions, including specific strategies for implementing them.

CALL TO ACTION 1: TRANSFORMATIVE CLIMATE LITERACY

Ensure that in all new basic and secondary education projects, basic climate literacy for all and transformative and gender-responsive E4CA are incorporated in the majority of curricula.

Key strategies include the following:

- Integrate education for climate action into national competencies frameworks and existing curricula across various subjects.
- Ensure that climate literacy learning outcomes include transformative competencies.
- Strengthen climate orientation of STEM education.
- Align teaching approaches with climate learning outcomes, such as place-based and action-oriented pedagogies.

CASE STUDY 1, IN SECTION 4, illustrates how investments into a holistic curriculum reform can promote climate literacy and resilience in secondary education in the Solomon Islands.

CALL TO ACTION 2: GREEN SKILLS

Ensure that in all new TVET, skills, and workforce development projects, green and climate resilience skills are introduced in occupational qualifications, linked to key economic sectors in a country's priority industries, and aligned with NDCs, national climate action plans, and economic development plans. Key strategies include the following:

- Update and/or introduce new green skills qualifications based on industry dialogue and green skills forecasting.
- Fund institutions to conduct green workforce forecasting, engage in industry dialog, and update qualification standards.
- Integrate basic environmental skills across all occupational standards.
- Strengthen and improve access to life-learning systems in coordination with active labor market programs.

CASE STUDY 2 is an example of how investments into in a state-of-the art skills university can champion education for climate action and workforce development in India. **CASE STUDY 3** shows how investments into specialized technology and training labs, including the introduction of training on sustainable business practices and eco-friendly teaching equipment, can improve the competitiveness of priority industries in Bangladesh.

CALL TO ACTION 3: EDUCATION FOR CLIMATE ACTION IN HIGHER EDUCATION

Ensure that in all new higher education projects, degree and post-graduate programs are updated or newly developed to include interdisciplinary climate knowledge and competencies in alignment with NDCs and national climate action plans. Key strategies include the following:

- Integrate education for climate action in degree programs across various disciplines using different integration approaches.
- Promote interdisciplinary degree programs and research.
- Develop new organizational structures in support of climate-oriented interdisciplinary teaching and research.
- Strengthen external partnerships and university alliances to access climate related knowledge and technologies.

CASE STUDY 4 shows how investments in a center for climate change education and green certification of faculty buildings can upgrade the quality and climate orientation of science and technology education and research in Sri Lanka.

CALL TO ACTION 4: CLIMATE ADAPTATION

Ensure that all new investments from 2025 in school and education institution infrastructure and operations, in any education subsector, are adapted to local climate and disasters risks and meet related building standards. Key strategies include the following:

- Assess climate risks and vulnerabilities of schools and develop, fund, and implement climate adaptation plans including structural and non-structural measures in all schools.
- Put in place social protection measures to address climate-related shocks to households with school children to prevent learning losses with a focus on disadvantaged households.
- Integrate climate resilience and disaster risk reduction into education sectors plans and budgets.

CASE STUDY 5 illustrates how investments into climate and disaster risk adaptation measures improve the climate resilience of schools transforming school safety and learning in Nepal.

CALL TO ACTION 5: CLIMATE MITIGATION

Ensure that all new investments from 2025 in school and education institution infrastructure, including newly built schools and renovations, are designed and constructed for resource efficiency, including through renewable energy sources, in line with national building codes and green building standards. Key strategies include the following:

- Conduct school energy audits and develop, fund and implement climate mitigation plans including structural and non-structural measures at schools.
- Build school staff capacity in sustainable school management.
- Integrate low-carbon school facilities into teaching and learning to facilitate place-based and inquiry-based learning.

CASE STUDY 6 shows how investments into climate mitigation measures can create a low-carbon school campus and industry-driven vocational education in the People's Republic of China.

CALL TO ACTION 6: CLIMATE INNOVATION

Ensure that in all new projects that support research, innovation, and start up incubation, the services of research and development and innovations hubs are climate-oriented, in alignment with NDCs, national climate action plans, and other economic and innovation strategies. Key strategies include the following:

- Build climate-oriented R&D, incubation, and commercialization capacities at universities and innovation hubs.
- Orient design of innovation hubs and its services toward the climate challenges and opportunities of local industries and communities.

CASE STUDY 7 illustrates how investments into science and technology parks can drive climate innovation in Indonesia.

CALL TO ACTION 7: NATIONALLY DETERMINED CONTRIBUTIONS

Ensure that all upcoming NDCs updates incorporate substantive measures and action plans for harnessing the role of education and skills for climate mitigation and adaptation through appropriate investments in human capital development. Key strategies include the following:

- Assess if education and skills development are substantially incorporated in NDCs and national climate action plans, and address identified gaps to make education part of climate action plans.
- Use the Joint MDB Methodological Principles for Assessment of Paris Agreement Alignment to guide the design of future education interventions and investments to ensure their Paris Agreement alignment.

CALL TO ACTION 8: MONITORING AND EVALUATION

Ensure that in all new education projects, measures and metrics are incorporated that track and monitor E4CA and related capacity and skills development of beneficiaries—including girls, women, and disadvantaged youth. Key strategies include the following:

- Develop metrics and indicators that reflect intended education for climate action outcomes.
- Integrate climate related education indicators and targets in strategies, plans, and design and monitoring frameworks.

Case Studies

Section 4 outlines seven case studies that illustrate how the eight calls to actions are already implemented in ADB-supported initiatives across Asia and the Pacific.

SOLOMON ISLANDS: SENIOR SECONDARY EDUCATION IMPROVEMENT PROJECT

Case Study 1

A HOLISTIC CURRICULUM REFORM TO PROMOTE CLIMATE LITERACY AND RESILIENCE IN SECONDARY EDUCATION IN THE SOLOMON ISLANDS

Project Context

The Solomon Islands face significant challenges as one of the most vulnerable nations globally to climate change and disasters, ranking second in disaster risk. The country frequently experiences extreme weather events and climate hazards (cyclones, flash floods, storm surges, king tides, and droughts), and geophysical hazards (volcanic activity, earthquakes, and tsunamis), which exacerbate poverty, gender inequality, and social vulnerability especially in rural areas and remote islands. The existing senior secondary curriculum, which has not been updated for over 30 years, does not equip students with the practical skills needed for a climate-resilient and gender-inclusive economy. Schools often lack climate-resilient infrastructure, with many buildings being unsafe, and lacking basic facilities. School management capacity needs to be upgraded to manage schools effectively, especially after disasters, and to promote gender-sensitive learning environments.

Activities to Promote Education for Climate Action

To address these challenges, the Government of the Solomon Islands, assisted by ADB, initiated the Senior Secondary Education Improvement Project to reform the country's senior secondary education.[9] At the heart of the project is the goal to prepare students to thrive in a climate-resilient economy by integrating climate, disaster resilience, and gender equality into every aspect of senior secondary curriculum. The project aims to align the education system with the National Development Strategy 2016–2035 and National Adaptation Programmes of Action.

---

[9]   ADB. Solomon Islands: Senior Secondary Education Improvement Project. The project is supported by grant funding from the Asian Development Fund and the Ireland Trust Fund for Building Climate Change and Disaster Resilience in Small Island Developing States.

The project adopts a multifaceted whole-of-school approach to greening senior secondary education and building resilience across multiple areas, including curriculum, teacher training, assessment methods, school leadership, and school infrastructure.

A central component of the project is the forthcoming extensive reform of the national curriculum for school years 10–12 that aims to make education for climate action a fundamental part of students' learning experience. The project will mainstream climate change education into 45 subject areas across the curriculum to equip students with the knowledge and breadth of skills needed to lead environmentally conscious lives, contribute to climate action and take up adaptive livelihoods and jobs in climate-vulnerable sectors, such as agriculture, fisheries, and tourism. For instance, students will learn about circular- and conservation-oriented economic models in economics classes; mathematical concepts using real-world scenarios related to climate change and disasters in mathematics classes; and climate-resilient approaches to agriculture and horticulture in agricultural science. The project will also support the introduction of a new core subject, "Living Sustainably," focusing on climate and disaster risk management, environmental advocacy, and sustainable living practices tailored to the Solomon Islands context.

Upgrades to teacher training will build the capacity of educators to effectively deliver the revised curriculum and foster a culture of resilience within schools. Teachers will receive professional development in the new and updated subjects including knowledge on climate and disaster resilience, climate-relevant pedagogies that reflect action-oriented and place-based learning, gender-inclusive education, and disability awareness. The training will focus on interdisciplinary approaches that will help teachers build students' transferable skills, such as critical thinking and problem solving. Moreover, the project will introduce and train teachers in new assessment frameworks aligned with new curricula.

School leadership will also be strengthened through targeted professional development programs that include disaster preparedness, climate change adaptation, and risk-informed infrastructure maintenance. School leaders will be trained in developing and implementing disaster response plans, coordinating with communities and youth, and helping to support schools to operate during disasters and avoiding prolonged periods of school closures. This training will include a transformative leadership and management program for female teachers, empowering them to become future school leaders and agents of change in their communities.

In addition, the project supports the upgrading of school facilities in 10 secondary schools tailored to the climate risk profile of each school to enhance schools' climate resilience and gender-responsiveness. The infrastructure improvements will include the construction and renovation of classrooms, dormitories, ablution blocks, and learning environments using sustainable design and practices, where possible. This includes climate resilience measures such as green building materials, low-carbon construction techniques, and passive cooling systems. The gender-responsive facility upgrading will improve students' safety, comfort, health, and include menstrual hygiene facilities and solar-powered lighting for enhanced security for girls and women.

The project also includes other learner-centered initiatives to target the immediate needs of beneficiaries. For example, a pilot program will be launched to support pregnant teens and young mothers in continuing their education; strengthen career guidance for climate-mitigative and climate-adaptive industries; encourage female participation in science, technology, engineering, and mathematics fields; and run a youth climate ambassador scheme—empowering students to take on leadership roles in their communities by promoting climate awareness and resilience.

In support of future actions, the project will help develop long-term infrastructure expansion plans and risk-informed maintenance strategies outlining additional climate adaptation and mitigation measures for funding. In addition, by updating and implementing national school infrastructure standards on the regulatory level, the project aims to set a new benchmark for climate-resilient educational facilities in Solomon Islands to develop the framework conditions for future climate-informed actions and investments.

Anticipated Outcomes

The anticipated outcomes of this project are far-reaching and transformative. By improving the quality, relevance, and gender-responsiveness of senior secondary education, the project will better prepare the next generation of Solomon islanders for a climate-resilient economy imparting students with the skills and knowledge necessary to pursue adaptive livelihoods and contribute to the sustainable development of their communities and the nation. Based on the holistic approach that includes reformed curriculum, including assessments, teacher training, upgraded climate-resilient infrastructure, and school management, the targeted schools are set to become climate-ready and will help to foster a culture of resilience, sustainability, and equality, laying the foundation for a more secure and prosperous future for all Solomon islanders. By fostering gender equality and social inclusion at the leadership level, the project aims to create a more equitable and resilient senior secondary education system.

INDIA: ASSAM SKILL UNIVERSITY PROJECT

Case Study 2

CHAMPIONING EDUCATION FOR CLIMATE ACTION AND WORKFORCE DEVELOPMENT THROUGH A STATE-OF-THE ART SKILLS UNIVERSITY IN INDIA

Project Context

Assam is the largest state in the northeastern region of India in terms of economy and population. It remains largely rural, with underdeveloped infrastructure and an economy reliant on low value-added, natural resource-based products. Its small manufacturing sector is undiversified and poorly integrated into economic value chains. Assam is also India's most climate-vulnerable state, facing extreme climate events such as floods, riverbank erosion, landslides, cyclonic storm (Assam State Action Plan on Climate Change 2015-2020). A key constraint to Assam's growth and climate resilience is the lack of a skilled, climate aware workforce essential for sustainable socioeconomic development.

To effectively address the state's human development needs, the Assam state government launched the Assam Skill University (ASU) project with ADB support.[10] The government envisions ASU as a hub for high-quality skills training, equipped with state-of-the-art facilities and technology, to improve employment and career development prospects by integrating skills training with higher education, applied R&D, entrepreneurship. ASU will boost the supply of skilled workers, enhancing the productivity and competitiveness of industries in Assam and the Northeastern Region.

---

[10]　ADB. India: Assam Skill University Project.

Activities to Promote Education for Climate Action

The project taking a whole-of-school approach supports the establishment of ASU by (i) developing university management and operating systems, business models, and faculties; (ii) designing industry-aligned, flexible skills education and training programs; (iii) building capacity to manage entrepreneurship, applied R&D, and technology transfer; (iv) improving access to professional development and skills training resources for other skills training institutions; and (v) constructing and operating an environmentally sustainable and climate-resilient campus.

The project supports climate adaptation and mitigation by developing environmentally sustainable, climate-resilient design and management systems for the university, including (i) green campus and building certification by the Indian Green Building Council; (ii) a building management system to monitor energy and water usage; (iii) disaster risk management and recovery plans for the main and interim campus; and (iv) training of university management, staff, and faculty on emergency protocols.

The project supports education for climate action (E4CA) through various approaches. It integrates climate literacy as a core skill by including a common climate change module in all undergraduate programs. This activity supports the 2021 Guidelines of the University Grants Commission that mandates environmental studies and climate change as compulsory subjects. These subjects cover climate change, sustainable development, biodiversity conservation, pollution control, sanitation, waste management, and forest and wildlife protection. It also integrates job-specific green skills into programs related to construction, forestry and biodiversity, and agriculture.

In addition, the project supports the design and delivery of labor market-relevant training and qualifications across a range of green and conventional industries, including (i) agricultural and food technology; (ii) technology; (iii) design and creativity; (iv) manufacturing and construction; (v) sustainability; (vi) mobility; (vii) management and finance; (viii) tourism, hospitality, and wellness; and (ix) health care. It emphasizes 21st century and entrepreneurial skills to prepare learners for self-employment and lifelong learning, given limited salaried job opportunities and rapidly changing skill requirements.

In addition, the project promotes green entrepreneurship by setting up and strengthening the capacity of a center for entrepreneurship and innovation, offering climate-oriented entrepreneurship programs and startup support.

Anticipated Outcomes

ASU is expected to become the premium institution for skills development in Assam becoming a leader in providing green skills and E4CA. It aims to service the broader education ecosystem in Assam as a hub for (i) quality assurance, (ii) capacity development for industrial training institutes and engineering colleges, (iii) business development for local industries, and (iv) socioeconomic development of the community. The anticipated outcomes include:

- ASU's environmentally sustainable, climate-resilient campus, recognized as a model for sustainability and resilience, offering education and training on environmental sustainability and climate change;
- integration of climate change modules into programs of other skills training institutions through professional development for teachers and faculty, curriculum development, and quality assurance provided by ASU's center for faculty and curriculum development;

- development of environmentally sustainable, climate-resilient businesses for local entrepreneurs, startups, and micro-, small-, and medium-sized enterprises through ASU's entrepreneurship and innovation center, offering skills training, R&D projects, technology transfer, and business development support; and
- enhanced climate and disaster resilience in the surrounding community through ASU's emergency shelter facilities and community outreach and awareness activities.

BANGLADESH: SKILLS FOR INDUSTRY COMPETITIVENESS AND INNOVATION PROGRAM

Case Study 3

IMPROVING THE COMPETITIVENESS OF PRIORITY INDUSTRIES THROUGH SPECIALIZED TECHNOLOGY AND TRAINING LABS IN BANGLADESH

Context

Bangladesh's coastal location and low-lying land render it one of the countries most vulnerable to climate change, leading to high economic and non-economic costs related to loss and damage (e.g., loss of life, mobility, environment, health, and/or knowledge). In the last decade, flooding incidences have worsened in terms of frequency and magnitude. Climate projections indicate that this trend is expected to continue with rising temperatures and precipitation. Failure to address climate change impacts is estimated to result in a 6.8% loss in GDP per year by 2030.

The increasing pressures of climate change, coupled with the global push for sustainability, demand that businesses shift toward climate-resilient and resource-efficient industry practices and adoption of green technologies essential for economic competitiveness. However, there is a significant gap in green skills among the current workforce and a shortage of competent mid-level managers, who can champion climate-resilient business models and environmentally conscious operations. These shortcomings hinder industries from integrating green technologies and sustainable practices that can increase industry competitiveness and move up the global value chains. Compounded by underemployment and rising youth unemployment, this lack of skilled green workforce contributes to a missed opportunity for inclusive green growth. The Mujib Climate Prosperity Plan 2030 highlights building a resilient and sustainable future by maximizing climate resilience and green opportunities, such as upskilling the workforce for the growth of quality green technology jobs.

Activities to Promote Education for Climate Action and Innovation

Responding to Bangladesh's high vulnerability to climate change and the opportunity to build a more resilient and sustainable future, the Skills for Industry Competitiveness and Innovation Program (SICIP) financed by ADB aims to drive the country's economic competitiveness and green growth through advanced technology-oriented skills development, including green skills.[11]

---

[11]   ADB. Bangladesh: Skills for Industry Competitiveness and Innovation Program.

The program is targeted at (i) increasing the technology-oriented skilled workforce across emerging and priority sectors; (ii) promoting inclusive skilling and upskilling opportunities, including green skills, for women and socially disadvantaged groups; (iii) incentivizing industry–university partnerships to nurture innovation capacity and improve industry competitiveness, with attention to green technologies and green skills; and (iv) fostering skills for climate-resilient manufacturing processes and green technologies.

The project supports various interventions for greening key economic sectors in Bangladesh, such as garments, textiles, leather goods and footwear, and light engineering. It promotes skills for climate-friendly manufacturing processes (e.g., sustainable waste management) with an added focus on developing competent mid-level managers, as these professionals are best positioned to lead and implement the adoption of advanced technologies and sustainable business operations in industries. It supports the establishment of smart textile technology labs at textile engineering colleges, which will follow green building design standards and be equipped with advanced energy and water-saving equipment in 35 practical training labs or workshops (e.g., procuring eco-friendly equipment). The improved facilities will provide access to practical training in advanced eco-friendly technologies, such as waterless dying technologies, and promote innovative green production processes in textile manufacturing. The SICIP will also provide a competitive research grant to foster industry-oriented research and development, prioritizing the application of digital and green technologies for industry solutions.

The program is expected to benefit about 220,000 new and existing workers, including women and people from socially disadvantaged groups, contributing to the creation of a technology-ready workforce. The workforce development activities will support the adoption of climate technologies in priority industries of Bangladesh to promote green growth and improved competitiveness in sustainable global value chains.

SRI LANKA: SCIENCE AND TECHNOLOGY HUMAN RESOURCE DEVELOPMENT PROJECT

Case Study 4

UPGRADING SCIENCE AND TECHNOLOGY EDUCATION THROUGH A CENTER FOR CLIMATE CHANGE EDUCATION, AND GREEN CERTIFIED FACULTY BUILDINGS IN SRI LANKA

Project Context

Sri Lanka has recorded steady economic growth of 5% throughout the 2010s. In light of rapid technological change, science and technology have been identified as critical to future economic development and to meet the human capital needs of key development initiatives such as Colombo–Trincomalee Economic Corridor. In addition, higher education learning opportunities in Sri Lanka have been limited, especially in the field of science and technology. The above average employment rate for science and technology graduates indicated an undersupply of this study field, amidst its increasing relevance for the transition to a low-carbon economy.

Activities to Promote Education for Climate Action

The Government of Sri Lanka initiated the Science and Technology Human Resource Development Project with the support of ADB.[12] The project aims to upgrade science and technology faculties of four universities with the goal to raise the quality of and expand the access to employment-oriented higher technology education and research including in the field of climate change. The project takes a holistic approach to improve the operations of four universities in key areas: (i) constructing modern faculty infrastructure, (ii) introducing new science and technology curricula and degree programs, (iii) establishing industry partnerships, and (iv) building human capacity of university faculty staff.

Based on climate change assessments, the project identified that selected university buildings were exposed to climate risks, such as heavy rainfall, storm surge and winds. Therefore, stakeholders recognized the need and opportunity to implement climate adaptation and mitigation measures to adapt building to climate risks to ensure learners and teachers' safety, improving the conditions of the learning environment and enhancing climate resilience of buildings.

The two constructed and upgraded faculty buildings—the Faculty of Computing and Technology Building at the University of Kelaniya and the Technology Faculty Building Complex at the Sabaragamuwa University of Sri Lanka—have already achieved a platinum certification under the Green Building Council of Sri Lanka; the highest classification of the national green building rating system. Climate adaptation and mitigation measures under the certification scheme encompasses eight key areas: (i) management; (ii) sustainable site; (iii) water efficiency; (iv) energy and atmosphere; (v) materials, resources, and waste management; (vi) indoor environmental quality; (vii) innovation and design process; and (viii) social and cultural awareness. The remaining faculty buildings under the project (i.e., Faculty of Engineering at the University of Sri Jayewardenepura and Faculty of Technology at the Rajarata University of Sri Lanka), which are currently under construction, also aim to be green certified upon completion. In addition, the project promotes the installation of rooftop solar panels at university to lower operational electricity costs and potentially generate revenue through energy sales to the main grid.

In support of E4CA and climate-oriented research, the project supports to establish the Center of Excellence in Climate Change Studies (CECCS) at the Rajarata University of Sri Lanka. CECCS aims to conduct multidisciplinary climate change related education, research, international collaboration, and awareness raising activities aimed at researchers, industry professionals and policymakers. The concept of the center was developed by an interdisciplinary group of university staff of different departments and outlines, among others, a shared vision and a dedicated management structure for the interdisciplinary center. During the initial stage, the center's activities will focus on providing skills and climate solutions to communities affected by climate change in dry agroecological zones.

For climate innovation, the project supports competitive research grants for engineering or technology faculty across all four universities, including research related to climate change. Examples of climate-oriented research grants include the development of a new real-time warning system for landslide prediction in Sri Lanka using Lidar remote sensing technology and international university partnerships to develop capacities for climate-smart agriculture.

---

[12] ADB. Sri Lanka: Science and Technology Human Resource Development Project.

Overall, the Science and Technology Human Resource Development Project supports the expansion and improvement of science and technology education to prepare students for skilled employment and accelerates the adoption of climate solutions in Sri Lanka through expanding research and development. The project also improves the climate resilience of facilities of four higher education institutions ensuring learning continuity and the well-being of learners. In line with the Paris Agreement, new faculty buildings were designed to achieve a platinum green building certification.

NEPAL: DISASTER RESILIENCE OF SCHOOLS PROJECT

Case Study 5

TRANSFORMING SCHOOL SAFETY AND LEARNING THROUGH CLIMATE RESILIENT SCHOOLS IN NEPAL

Project Context

As climate-induced disasters increase in frequency and severity, Nepal's students face escalating threats to their safety, learning environments, and education. This growing danger underscores the urgent need for resilient school infrastructure that can withstand and recover from such conditions. Nepal's educational infrastructure is acutely vulnerable to disasters, including earthquakes, landslides, floods, extreme heat, and drought. The 2015 earthquake, which caused $7 billion in damage, injured over 22,000 people, and damaged more than 7,800 schools, highlights the urgent need for safer, resilient educational facilities. Many schools still operate below modern safety standards, with 3,569 partially damaged schools posing ongoing risks to students and staff. As climate change exacerbates these hazards, building resilient schools is imperative to protect schoolchildren and to ensure the continuity of education in Nepal.

Project Activities for Climate Resilient Schools

Through the Disaster Resilience of Schools (DRS) project, ADB supports the reconstructing and retrofitting of 163 public schools across three provinces in Nepal, including 145 secondary schools and 18 feeder primary schools.[13] The project focuses on 14 districts severely impacted by the 2015 earthquake, ensuring that these schools meet earthquake and climate-related standards, offering safe, sustainable learning environments. Upgraded facilities include science labs, information and communication technology (ICT)-equipped rooms, libraries, and improved sanitation.

The project aligns with the Comprehensive School Safety Framework, emphasizing safe and resilient infrastructure and learning facilities built with green development concepts, school disaster risk management (DRM) planning and DRM education. It prioritizes climate risk reduction and mitigation, climate adaptation, and environmental protection throughout its design and implementation. For example, enhancements to the Education Management Information System help to assess infrastructure needs and involve communities in disaster risk management, with a focus on gender-sensitive and inclusive planning. Local governments and educational authorities have benefited from capacity building efforts on disaster

---

13   ADB. Nepal: Disaster Resilience of Schools Project.

risk reduction planning, and government engineers and masons have been trained in disaster-resilient construction techniques.

Community-based retrofitting initiatives were also introduced to boost disaster resilience, using local funding for scalable, replicable approaches. Solar backup systems and high-grade solar plants were installed in 130 schools, reducing reliance on fuel generators and supporting climate adaptation. These solar systems provide a reliable and stable power supply for ICT rooms, libraries, and other essential facilities, particularly during outages or emergencies, while contributing to environmental sustainability by decreasing reliance on non-renewable energy. Training programs for solar electric technicians were included to ensure local capacity for maintaining and operating these systems, supporting the project's long-term sustainability.

Outcomes

The DRS project made substantial strides in comprehensively improving the safety and resilience of schools impacted by the 2015 earthquake. Covering 346 schools, the project reconstructed 205 heavily damaged schools and retrofitted 141 schools to meet earthquake-resistant standards, with 84 schools receiving both reconstruction and retrofitting. Schools were equipped with ramps, disability-friendly water, sanitation and hygiene facilities, gender-separated restrooms, and solar backup systems and higher-grade solar plants to reduce the reliance on fuel generators. Recreational facilities, emergency preparedness plans, and evacuation procedures were also introduced. Climate risk mitigation and adaptation training programs educate students on responding to emergencies such as heavy rainfall, high winds, landslides, floods, heatwaves, and epidemics.

PEOPLE'S REPUBLIC OF CHINA: GUANGXI BAISE VOCATIONAL EDUCATION DEVELOPMENT PROJECT

Case Study 6

INDUSTRY-RESPONSIVE VOCATIONAL EDUCATION ON A LOW-CARBON UNIVERSITY CAMPUS IN THE PEOPLE'S REPUBLIC OF CHINA

Project Context

After experiencing slow growth for decades, Guangxi Zhuang Autonomous Region (GZAR) has prospered under the People's Republic of China (PRC) strategy for development of its western region. The city of Baise is being developed into a new economic base in the GZAR by focusing on the development of higher-value-chain production in four priority industries: (i) aluminum processing, (ii) agriculture, (iii) tourism, and (iv) regional trade and logistics. The surge in economic growth, large-scale rural–urban migration, and industrial restructuring resulted in the need for a local skilled workforce and ensuring sustainable development enabling a safe, healthy, and livable environment. To meet labor market needs traditionally academic universities aimed to transform into more market-responsive TVET institutions that promote pathways between TVET, higher education, and adult education, and facilitate collaboration between industry and education. In addition, the *National Action Plan for Energy Conservation, Emissions Reduction, and Low Carbon Development* set key directions for universities, colleges, and public institutions to undertake energy saving and emission reduction.

Activities to Promote Sustainable and Low-Carbon School Facilities

In response these developments, the ADB-financed Guangxi Baise Vocational Education Development Project supported Baise University in developing a multi-level TVET system establishing connections between tiered training programs to promote continuous skills development opportunities, pathways between vocational and academic education and an improved transition into careers.[14] The project supported purpose-built infrastructure at the Chengbi campus and provided specialized equipment, curriculum development, and teacher capacity building to enable competency-based learning across TVET levels in key sectors, such as food science and engineering.

Project activities were informed by considerations for sustainability, resource-efficiency and low-carbon development. It supported the development of an environmentally sustainable campus through the establishment of the Green Sustainability Center mandated to oversee the campus sustainability strategy. This strategy introduced three sustainability pillars for the operations of Baise University: (i) green logistics and general services, (ii) green curriculum, and (iii) green community, thereby taking a holistic approach. Under the first pillar, a Water and Electricity Management Center was created to manage new amenities like electric vehicle charging stations and monitor the usage of water and electricity of the campus in accordance with the Energy Conservation Staff Manual set out by the government. Under the second pillar, faculty from the Chemistry and Environmental Engineering departments integrated sustainable development into the curriculum, focusing on operations of reclaimed water stations, e-waste recycling, and research into using biological resources such as plant and organic waste to create sustainable energy. Green teaching concepts were also integrated into TVET capacity development workshops. The third pillar promoted public participation in green awareness through student organizations such as the Student Energy Conservation and Environmental Protection Association. Activities like speaker events, competitions, and tree planting created place-based and community-centered learning opportunities, key teaching methods of E4CA.

With a campus-wide focus on sustainability, Baise University leverages facilities as a living learning lab providing students with opportunities for real-world problem solving. For instance, students engaged with the Green Sustainability Center to address water management issues on campus by conducting compliance monitoring of the on-campus sewage treatment stations.

Other initiatives advancing the campus sustainability strategy included the design of new buildings in compliance with green and energy-efficient codes of the provincial government, installation of high-efficiency heat pumps for air conditioning and hot water in dormitories, and a 3.0 megawatt (MW) photovoltaic power system, generating capacity of 3.879 MW-hour annually, significantly contributing to the campus's electricity needs. By project completion, Baise University's renewable energy consumption accounted for more than 18%, surpassing the 15% target. Advances in equipment technology enabled a higher-than-expected share of renewable energy use.

The Guangxi Baise Vocational Education Development Project is strategically aligned with government action plans and building regulations. It combined activities aimed at improving education quality and relevance with climate-specific measures across teaching and learning, campus management, and community engagement.

---

[14]   ADB. China, People's Republic of: Guangxi Baise Vocational Education Development Project.

INDONESIA: PROMOTING RESEARCH AND INNOVATION THROUGH MODERN AND EFFICIENT SCIENCE AND TECHNOLOGY PARKS PROJECT

Case Study 7

DRIVING CLIMATE INNOVATION THROUGH SCIENCE AND TECHNOLOGY PARKS IN INDONESIA

Project Context

Indonesia has sustained average economic growth rates above 5% since 2000 and has made significant progress in reducing poverty. As a middle-income country, challenges to future economic growth include Indonesia's slowing productivity growth, competitiveness issues associated with global technological transformation, and concerns over sharp increases in its greenhouse gas (GHG) emissions. At the same time opportunities are emerging. Technology adoption could add up to $2.8 trillion to the Indonesian economy by 2040, spurring growth in GDP by an additional 0.55 percentage points annually over the next 2 decades.[15]

To act upon the opportunities of tomorrow, the Ministry of Education, Culture, Research and Technology with the support of ADB has launched the Promoting Research and Innovation through Modern and Efficient Science and Technology Parks (PRIMESTeP) Project.[16] PRIMESTeP aims at supporting R&D, commercialization, and startup incubation with a focus on climate mitigation and adaptation solutions to incentivize and promote innovation and investment that can lead to long-term environmental and economic benefits.

Activities to Promote Climate Innovation

The PRIMESTeP Project focuses on building the capacity of four strategically selected science and technology parks (STPs) at the Bandung Institute of Technology, Gadjah Mada University, IPB University, and University of Indonesia. These STPs are based at Indonesia's top autonomous public research-intensive universities and align with Indonesia's priority economic sectors.

PRIMESTeP strengthens these STPs in three critical areas—facilities, human capacity, and innovation funding. It will upgrade advanced research and development labs, equip them with Fourth Industrial Revolution technologies, and provide $30 million in grants to applied research teams for community, commercial or industrial-focused R&D. Another $10 million will support startup incubation for students, faculty, and alumni to help turn incubated companies into financially viable enterprises.

The project will also enhance researcher capabilities, especially for female researchers, through international post-doctorate scholarships in priority sectors. STP administrators will be trained to develop, implement, and monitor innovation partnerships between academia, industry, and government. Additionally, the Ministry of Education, Culture, Research and Technology and STP officials will receive training in financial management, procurement, gender mainstreaming, and safeguards monitoring.

---

[15] ADB. 2020. Innovate Indonesia: Unlocking Growth Through Technological Transformation. March.

[16] ADB. Indonesia: Promoting Research and Innovation through Modern and Efficient Science and Technology Parks Project.

PRIMESTeP promotes multi-stakeholder engagement by forging networks and partnerships. These connect (i) applied researchers with technology readiness levels (TRL) 5–9; (ii) industry partners seeking innovative solutions or commercialization opportunities that align with the project priorities; (iii) the university staff, alumni, and student community seeking entrepreneurship support; and (iv) leaders responsible for institutional transformation.

Research and incubation activities are strategically aligned with Indonesia's green growth road map, low-carbon development initiative, and updated nationally determined contributions. Climate mitigation R&D focuses on vehicle electrification, biomass from agriculture by-products, energy-efficient building technology, and energy storage technology. Climate adaptation R&D projects focus on climate-smart agriculture, disaster mitigation, and sustainable food production.

Anticipated Outcomes

The project aims to enhance Indonesia's economic competitiveness and drive sustainable growth by improving the quality and relevance of R&D and innovation at four STPs. It is expected to produce 167 innovations or products; 18 strategic R&D projects with a focus on social inclusion, gender mainstreaming, or climate; and 21 joint research projects. Through the startup incubation grant, 3,750 students will receive entrepreneurship training, and 470 startup teams will be incubated.

PRIMESTeP is making significant progress in its second year. Climate-related R&D and startup acceleration projects in fisheries, sustainable agriculture, circular waste management, biofuels, and battery efficiency are advancing toward commercialization. These projects underscore the importance of climate innovation investments in R&D and startup to accelerate sustainable economic development in emerging economies. Examples of current R&D projects include:

(i)   Bioconversion of organic waste (Biomagg). A start-up specializing in industrial-scale organic waste management using bioconversion technology, Biomagg transforms organic waste into alternative protein sources for animal feed and organic fertilizers, contributing to environmental sustainability.

(ii) Smart and resource-efficient irrigation. A startup developing an Internet of Things-based smart irrigation system that allows farmers to monitor and control irrigation remotely, optimizing water usage and improving agricultural productivity through their smartphones.

(iii) Development of New Generation Battery Components. Research on improving lithium-ion battery cathodes ($LiNi0.5Mn1.5O4$), via solid-state synthesis, aims to enhance electric vehicle battery cycle performance, safety, and charging speed. Initial results show promising cycle stability and capacity retention over 250 cycles.

Bibliography

Asian Development Bank (ADB). 2015. *University - Industry - Government Partnership for Economic Development in Indonesia.*

———. 2021a. The Female Secondary Stipend and Assistance Program in Bangladesh What Did It Accomplish? *ADB South Asia Working Paper Series.* No. 81.

———. 2021b. *Incubating Indonesia's Young Entrepreneurs Recommendations for Improving Development Programs.*

———. 2023a. *Asia in the global transition to net zero: Asian Development Outlook 2023 Thematic Report.*

———. 2023b. *Preparing the workforce for the low-carbon economy: A closer look at green jobs and green skills.*

———. 2023c. *Climate Change Action Plan, 2023–2030.*

———. 2024a. *Data Stories. Under SDG 13, Climate Action is Regressing Alarmingly in Asia and the Pacific.*

———. 2024b. *ADB is Asia and the Pacific's Climate Bank.*

———. 2024c. Handbook on Energy Efficiency in Buildings.

———. 2024d. Asia-Pacific Climate Report 2024.

ADB et al. 2023a. *Joint MDB methodological principles for assessment of Paris Agreement alignment of new operations: Direct investment lending operations.*

———. 2023b. *Common Principles for Climate Mitigation Finance Tracking.*

Accenture. 2022. *Youthquake Meets Green Economy—Why Businesses Need to Care.*

J. Affolderbach and R. Krueger. 2017. "Just" ecopreneurs: Re-conceptualising green transitions and entrepreneurship. *Local Environment.* 22 (4). pp. 410-423.

J.C. Allen and S. Malin. 2008. Green entrepreneurship: A method for managing natural resources? *Society & Natural Resources.* 21 (9). pp. 828-844.

N. Angrist et al. 2024. Human Capital and Climate Change. *New Working Paper Series.* Working Paper 31000. National Bureau of Economic Research.

N. Ardoin, A.W. Bowers, and E. Gaillard. 2020. Environmental education outcomes for conservation: A systematic review. *Biological Conservation.* 241. 108224.

M.N. Asadullah, K.M.M. Islam, and Z. Wahhaj. 2021. Child marriage, climate vulnerability and natural disasters in coastal Bangladesh. *Journal of Biosocial Science.* 53 (6). pp. 948-967.

M. Bacigalupo et al. 2016. *EntreComp: The Entrepreneurship Competencies Framework.* Publications Office of the European Union.

Banco Bilbao Vizcaya Argentaria (BBVA) Research. 2024. *Economics of Climate Change: Green Innovation to Boost Activity and Cut Emissions.* Weekly Summary 1 March.

G. Bianchi, U. Pisiotis, and M. Cabrera Giraldez. 2022. *GreenComp: The European sustainability competencies framework.*

J. Bentz and K. O'Brien. 2019. Art for change: Transformative learning and youth empowerment in a changing climate. *Elementa: Science of the Anthropocene* 7: 52

A. de Bruin. 2016. Towards a framework for understanding transitional green entrepreneurship. *Small Enterprise Research.* 23 (1). pp. 10-21.

K. Brundiers et al. 2020. Key comptencies in sustainability in higher education--toward and agreed--upon reference framework. *Sustainability Science.* 16. pp. 13-29.

K. Brundiers, J. King, R. Parnell, and K. Hiser. 2023. A GCSE proposal statement on key competencies in sustainability: Guidance on the accreditation of sustainability and sustainability-related programs in higher education. Global Council for Science and the Environment.

IFHV (Bündnis Entwicklung Hilft). 2023. *World Risk Report 2023.*

Bureau of Energy Efficiency – India. 2009. *Energy Management in Your School.*

CEDEFOP and OECD. 2022. Apprenticeships for greener economies and societies.

A. Chaturvedi et al. 2024. *Skill gap assessment across green hydrogen sector in India: Final report.* South Asia Regional Energy Partnership (SAREP).

B. Clarysse et al. 2005. Spinning Out New Ventures: A Typology of Incubation Strategies from European Research Institutions. *Journal of Business Venturing.* 20 (2). pp. 183–216.

Climate and Industry Research Team. 2023. *Advancing Climate Literacy in Union Vocational Education and Training Programs in English Canada, Quebec, Europe and the US: Analysis, Findings and Lessons Learned.*

Common Worlds Research Collective. 2020. Learning to become with the world: Education for future survival. Background paper for the Futures of Education initiative. UNESCO.

E. Dabla-Norris et al. 2023. *Asia's perspectives on climate change: Policies, perceptions, and gaps.* International Monetary Fund.

C.C. David et al. 2018. School hazard vulnerability and student learning. *International Journal of Educational Research* 92. pp. 20-29.

Environment Protection Department – Hong Kong. 2024. *Guide to Low Carbon Schools. Practical Guide on Carbon Audit and Management.*

European Investment Bank. 2022. *Extract of the 2021 Joint Report on Multilateral Development Banks' Climate Finance – Joint Methodology for Tracking Climate Change Adaptation Finance.*

European Training Foundation (ETF). 2024. *The Future of Skills in ETF Partner Countries.*

ETF, European Centre for the Development of Vocational Training, and the International Labour Office. 2016. *Developing Skills Foresights, Scenarios and Forecasts. Guide To Anticipating and Matching Skills and Jobs* (Volume 2).

ETF and UNICEF. 2021. *Building a Resilient Generation in Central Asia and Europe—Youth Views on Lifelong Learning, Inclusion, and the Green Transition.*

European Commission. 2022. *Learning for the Green Transition and Sustainable Development* - Publications Office of the EU.

L. Eusterbrock. 2024. Eco hip-hop education: Rap music, environmental justice, and climate change in music education. In L. Eusterbrock, C. Kattenbeck, and O. Kautny (Eds.). It's how you flip it: Multiple perspectives on hip-hop and music education. pp. 293. Transcript.

T. Garg, M. Jagnani, and V. Taraz. 2020. Temperature and human capital in India. *Journal of the Association of Environmental and Resource Economics.* 7(6).

GIZ (Deutsche Gesellschaft für Internationale Zusammenarbeit). 2022. *Energy Efficiency in the Building Sector for a More Sustainable Mongolia.*

Global Education Monitoring (GEM) Report and Monitoring and Evaluating Climate Communication and Education (MECCE) Project. 2024. Education and climate change: Learning to act for people and planet. UNESCO.

Government of the Philippines, Department of Public Works and Highways. 2015. The Philippine Green Building Code.

Growing Green. 2024. Growing green framework.

A. Hindley. 2022. Understanding the Gap between University Ambitions to Teach and Deliver Climate Change Education. *Sustainability.* 14 (21). p. 13823.

HKPC. n.d. Guide to low carbon schools: Practical guide on carbon audit and management.

Hochschule für Agrar- und Umweltpädagogik . 2018. *Green pedagogy: From theoretical basics to practical sustainable learning activities.*

J. Holst, J. Grund, and A. Brock. 2024. Whole Institution Approach: Measurable and highly effective in empowering learners and educators for sustainability. *Sustainability Science.* 19. pp. 1359-1376.

Inner Development Goals. n.d. Inner Development Goals Framework.

International Climate Initiative and UNEP. 2023. *Integrating Electric 2&3 Wheelers into Existing Urban Transport Modes in Developing and Transitional Countries – Vietnam.*

International Energy Agency (IEA). 2022a. *World Energy Employment.*

————. 2022b. Reports – *Understanding Gender Gaps in Wages, Employment and Career Trajectories in the Energy Sector.*

————. 2023. *World Energy Employment 2023.* World_Energy_Employment_2023.pdf (iea.blob.core. windows.net)

International Labour Organization (ILO). 2021. *Global Framework on Core Skills for Life and Work in the 21st Century.*

————. 2024. *Climate Change and Financing a Just Transition.*

International Union for Conservation of Nature (IUCN). 2011. *Terminologies Used in Climate Change.*

B.A. Janney, L. Zummo, and M. Lohani. 2024. Shifting the desired outcome from climate literacy to climate agency: Education that empowers civic leaders. *Interdisciplinary Journal of Environmental and Science Education.* 20 (3). e2412.

V. M. Joffre and P. Serafica. 2023. Women Must Be at The Forefront of The Transition to A Low-Carbon Economy. Asian Development Blog. 29 March.

A. Kalpaka, J. Sörvik, and A. Tasigiorgou. 2020. *Digital Innovation Hubs as Policy Instruments to Boost Digitalisation of SMEs.* Publications Office of the European Union.

J. Kastrup and W. Kuhlmeier. n.d. Bestimmung und Beschreibung nachhaltigkeitsorientierter Kompetenzen in der Beruflichen Bildung. BBNE für Berufsbildung.

M. Khanna and N. Kochhar. 2023. Do marriage markets respond to a natural disaster? The impact of flooding of the Kosi river in India. *Journal of Population Economics.* 36. pp. 2241-2276.

C. Kwauk. 2021. *Who's Making the Grade on Climate Change Education Ambition?* Commentary Op-Ed. The Brookings Institution.

C. Kwauk et al. 2022. Education. In *State and Trends in Adaptation Report 2022.* Global Center on Adaptation.

C. Kwauk and O. Casey. 2022. A green skills framework for climate action, climate empowerment, and climate justice. *Development Policy Review.* 40 (S2): e12624.

J. Lassa, M. Petal, and A. Surjan. 2022. Understanding the impacts of floods on learning quality, school facilities, and educational recovery in Indonesia. *Disasters.* 47 (2). pp. 412–436.

A. Lavopa (United Nations Industrial Development Organization [UNIDO]) and M. Menéndez (UNU-MERIT). 2023. *Who is at the forefront of the green technology frontier? Again, it's the manufacturing sector.* UNIDO.

Leiden-Delft-Erasmus Universitis. n.d. LDE Centres and Programmes.

LinkedIn. 2023. *Global Green Skills Report 2023.*

MECCE and NAAEE. 2022. Mapping the landscape of K–12 climate change education policy in the United States.

Ministry of Planning Development and Special Initiatives. 2022.Pakistan Floods 2022: Post-Disaster Needs Assessment.

M. Monroe et al. 2019. Identifying effective climate change education strategies: A systematic review of the research. *Environmental Education Research.* 25 (6). pp. 791-812.

C.V. Nguyen and N.M. Pham. 2018. The impact of natural disasters on children's education: Comparative evidence from Ethiopia, India, Peru, and Vietnam. *Review of Development Economics.* 22 (4). pp. 1561–1589.

Organisation for Economic Co-operation and Development (OECD). 2023. PISA 2025 Science Framework (Draft).

M. Pavlova and C.S. Chen. 2019. Facilitating the development of students' generic green skills in TVET: An ESD pedagogical model. TVET@Asia, 12.

M. Pelli and G Tschopp. 2024. Storms, Early Education, and Human Capital. *ADB Economics Working Paper Series.* No. 743. Asian Development Bank.

L. Phelan et al. 2015. *Learning and teaching academic standards statement for environment and sustainability.* Office for Learning and Teaching.

Philippine Institute for Development Studies (PIDS). 2024. *A month of teaching days lost due to extreme heat, other calamities: Study.*

Programme for Energy Efficiency in Buildings (PEEB). 2023. Buildings in the NDCs: Mapping targets on buildings in the Nationally Determined Contributions.

C.M. Prentice et al. 2024. Education outcomes in the era of global climate change. *Nature Climate Change.* 14. pp. 214–224.

C. Preston. 2024. How colleges can become 'living labs' for combating climate change. Hechinger Report and Grist.

G. PV and C. Subudhi. 2020. Psycho-social and educational impact of flood among school going children. *Journal of Social Welfare and Management.* 12 (1).

F.M. Reimers. The Role of Universities Building an Ecosystem of Climate Change Education. *Education and Climate Change.* 2020 (4). pp. 1–44.

Z.P. Robinson et al. 2021. Universities as living labs for climate praxis. In. C. Howarth, M. Lane, and A. Slevin (Eds.). *Addressing the climate crisis.* pp. 129-139. Palgrave Macmillan.

D. Rousell and A. Cutter-Mackenzie-Knowles. 2020. A systematic review of climate change education: Giving children and young people a 'voice' and a 'hand' in redressing climate change. *Children's Geographies.* 18 (2). pp. 191-208.

J. Saavedra and L. Serburne-Benz. 2022. *Pakistan's floods are deepening its learning crisis.* World Bank.

D. Sarpong et al. 2023. The Three Pointers of Research and Development (R&D) for Growth-Boosting Sustainable Innovation System. *Technovation.* 122. p. 102581.

N. Schady et al Forthcoming. Heat and Learning: How Exposure to Extreme Heat Affects Learning in Brazil.

J. Schot and W. E. Steinmueller. 2018. Three frames for innovation policy: R&D, systems of innovation and transformative change. *Research Policy.* 47 (9). pp. 1554–1567.

M. Shah and B.M. Steinberg. 2017. Drought of opportunities: Contemporaneous and long-term impacts of rainfall shocks on human capital. *Journal of Political Economy.* 125 (2). pp. 527-561.

Skill Council for Green Jobs. 2023. *Gearing Up the Indian Workforce for a Green Economy: Mapping Skills Landscape for Green Jobs in India.*

————. 2024. *Green Jobs Handbook.*

A. Schröder. 2023. *Blueprint for an Industry Driven Long-Term Skills Strategy.* Presentation for European Steel Skills Alliance: Final Conference. Thyssenkrupp Steel Europe. 11 May 2023.

C. Sorensen et al. 2018. Climate Change and Women's Health: Impacts and Policy Directions. *PLoS Medicine.* 15 (7).

G. Subrahmanyam. 2020. *Future of TVET teaching.*

Sustainability Victoria. 2016. Energy: A 'how to' guide.

Sustainable Development Solutions Network (SDSN). 2022. Net zero on campus: A guide for universities and colleges to accelerate climate action.

Temasek and Ecosperity. 2021. *New nature economy: Asia's next wave. Risks, opportunities, and financing for a nature-positive economy.* World Economic Forum and AlphaBeta.

W. Thiery et al. 2021. Intergenerational inequities in exposure to climate extremes: Young generations are severely threatened by climate change. *Science.* 374 (6564). pp. 158–160.

A. Tyagi et al. 2022. *India's Expanding Clean Energy Workforce.* Council on Energy, Environment and Water, Natural Resources Defense Council, and Skill Council for Green Jobs.

United Nations Environment Programme (UNEP). 2021. *A practical guide to climate-resilient buildings and communities.*

United Nations Development Programme (UNDP). 2023. *The Climate Dictionary.*

———. 2024. *Women in Science, Technology, Engineering and Mathematics (STEM) in the Asia Pacific.*

United Nations Educational, Scientific and Cultural Organization (UNESCO). 2021. *Getting every school climate-ready: How countries are integrating climate change issues in education.*

———. 2022. *Youth demands for quality climate change education.*

———. 2023a. *Asia-Pacific regional synthesis: Climate change, displacement and the right to education.*

———. 2024a. *Green school quality standard: Greening every learning environment.*

———. 2024b. *Greening curriculum guidance: Teaching and learning for climate action.*

———. n.d. *Climate Change Education.*

UNESCO and Education International. 2021. *Teachers have their say: Motivation, skills and opportunities to teach education for sustainable development and global citizenship.*

UNESCO IBE. 2017. *Developing and implementing curriculum frameworks.*

UNESCO Institute for Statistics. n.d. *Education for Sustainable Development.*

UNESCO International Institute for Education Planning. n.d. *Gender-Responsive Pedagogy.*

UNESCO-UNEVOC. 2017. *Greening technical and vocational education and training: A practical guide for institutions.*

———. 2021. Skills development and climate change action plans. Enhancing TVET's contribution.

United Nations Framework Convention on Climate Change (UNFCCC). 2023. *Declaration on the common agenda for education and climate change at COP28.*

United Nations Children's Fund (UNICEF). 2021. *The climate crisis is a child rights crisis: Introducing the Children's Climate Risk Index.*

———. 2024. *Skills for a green transition: Solutions for youth on the move.* ILO, World Bank, UNICEF.

United Nations Conference on Trade and Development. 2019. *A Framework for Science, Technology and Innovation Policy Reviews Harnessing Innovation for Sustainable Development.*

United Nations Economic and Social Commission for Asia and the Pacific (UNESCAP). 2023. *Asia-Pacific Population and Development Report 2023 (ST/ESCAP/3112).*

US Global Change Research Program. 2024. *Climate Literacy: Essential Principles for Understanding and Addressing Climate Change.*

S. Venegas Marin, L.N.T., Schwartz, and S. Sabarwal. 2024. *The impact of climate change on education and what to do about it.* World Bank.

Vivid Economics. 2021. *Skills for the low carbon transition.*

K. S. Wangda and W. Phodrang. 2024. TTI-Samthang Adopts Simulators for Safer and Cost-Effective Learning. *BBS.bt.* 5 September.

World Bank. n.d. *Global Program for Safer Schools Global Library of School Infrastructure.*

World Economic Forum. 2019. *Systems Leadership Can Change the World – But What Exactly Is It?*

———. 2020. *Stronger schools and brighter futures in Tonga.* World Bank.

Young Lives. 2022. *Protecting the Most Vulnerable People in Vietnam from Climate Shocks: Exposure to Flooding and Crop Failures has Unequal Impact on Children's Development and Learning.*

www.ingramcontent.com/pod-product-compliance
Lightning Source LLC
LaVergne TN
LVHW071451180726
843512LV00018B/1348

9 789292 709907